Nimble with Numbers

Engaging Math Experiences to Enhance Number Sense and Promote Practice

Grades 1 and 2

Leigh Childs, Laura Choate, and Karen Jenkins

Dale Seymour Publications
Parsippany, New Jersey

Thanks to special friends and colleagues, who helped field test these activities with their students and who gave valuable feedback and suggestions:

Sandy Norton

Sandy Jenkins

Thanks to super kids who provided a student perspective as they tried these activities:

Children in Room 11 at Maie Ellis Elementary School, Fallbrook, California

Rhody Mashek

Managing Editor: Catherine Anderson
Senior Editor: Carol Zacny
Consulting Editor: Dorothy Murray
Production/Manufacturing Director: Janet Yearian
Production/Manufacturing Manager: Karen Edmonds
Production/Manufacturing Coordinator: Joe Conte
Design Manager: Jeff Kelly
Text design: Tani Hasegawa
Page composition: Susan Cronin-Paris
Cover design: Ray Godfrey
Art: Stefani Sadler

Dale Seymour Publications
An imprint of Pearson Learning
299 Jefferson Road
Parsippany, NJ 07054
www.pearsonlearning.com
1-800-321-3106

ISBN 0-7690-2721-0
4 5 6 7 8 9 10-ML-05-04-03-02-01-00

Table of Contents

Place Value

Money

Blackline Masters

Introduction to *Nimble with Numbers*

Why This Book?

National recommendations for a meaning-centered, problem-solving mathematics curriculum place new demands on teachers, children, and parents. Children need a facility with numbers and operations to achieve success in today's mathematics programs. Basics for children today require a broadening of the curriculum to include all areas of mathematics. Children are being asked to demonstrate proficiency not just in skills, but in problem solving, critical thinking, conceptual understanding, and performance tasks. Consequently, the reduced time teachers devote to number must be thoughtful, selective, and efficient.

This book fulfills the need for high-quality, engaging math experiences that provide meaningful practice and further the development of number sense and operation sense. **These activities are designed to help children practice number concepts previously taught for understanding in a variety of contexts.** Besides meeting the need for effective practice, *Nimble with Numbers:*

- provides a variety of adaptable formats for essential practice;
- supplements and enhances homework assignments;
- encourages parent involvement in improving their child's proficiency with basic facts and number concepts; and
- provides motivating and meaningful lessons for a substitute teacher.

Criteria for Preferred Activities

For efficient use of time devoted to the number strand, the book focuses on activities that are:

- Inviting (encourages participation)
- Engaging (maintains interest)
- Simple to learn
- Repeatable (able to reuse often and sustain interest)
- Open-ended, allowing multiple approaches and solutions
- Easy to prepare
- Easy to adapt for various levels
- Easy to vary for extended use

In addition, these activities:

- Require a problem-solving approach
- Improve basic skills
- Enhance number sense and operation sense
- Encourage strategic thinking
- Promote mathematical communication
- Promote positive attitudes toward mathematics as mathematical abilities improve

Planning Made Simple

Organization of the Book

The activities of this book are divided into six sections that cover high-priority number topics for first graders and second graders. The first section provides practice with counting and comparing numbers through 30 and identifying one and two more or less. The next three sections review the addition and subtraction facts. Since *ten* plays such an important role in our numeration system, many activities in Addition Facts, Subtraction Facts, and Mixed Facts emphasize ten as an important anchor and serve as a good lead-in to the Place Value Section.

The last section reinforces money concepts which gives children an opportunity to practice real-life skills. Throughout all sections, we make an obvious attempt to promote number sense.

Each section begins with an overview and suggestions to highlight the activities and provide some timesaving advice. The interactive activities identify the specific topic practiced (Topic), the objective (Object), the preferred grouping of participants (Groups), and the materials required (Materials). At the end of activities, "Making Connections" questions promote reflection and help children make mathematical connections. Tips are included to provide helpful implementation suggestions and variations. Needed blackline masters are included with the activity or in the Blackline Masters section at the end of the book.

This introductory section includes a Matrix of Activities. The repeatable Sponges and Games are listed alphabetically with corresponding information to facilitate their use.

Types of Activities

The book contains activities for whole group, small groups, pairs, and individuals. Most sections provide:

- Sponges (S)
- Skill Checks (C)
- Games (G)
- Independent Activities (I)

Sponges

Sponges are enriching activities for soaking up spare moments. Use Sponges with the whole class or with small groups as warm-up activities, or during spare time to provide additional math practice. Sponges usually require little or no preparation and are short in duration (3–15 minutes). These appealing Sponges are repeatable and, once they become familiar, can be student-led. Children are motivated to finish a task when they know a favorite Sponge will follow.

Skill Checks

The Skill Checks in each section provide a way to show children's improvement to parents as well as to themselves. Each page is designed to be duplicated and cut in half, providing six comparative records for each child. Children should begin by responding to the starter task following the STOP sign. These starter tasks are intended to promote number sense. Some teachers believe children perform better on the Skill Checks if the responses to the STOP task are shared and discussed before children solve the remaining problems. Most children will complete a Skill Check in 10 to 15 minutes. The concluding extension problem, labeled "Go On," accommodates those children who finish early. We recommend that early finishers be encouraged to create similar problems for others to solve. By having children discuss their approaches and responses, teachers help children discover more efficient number strategies.

Games

Even though Games are intended to be played by pair players or pairs, a new Game might be introduced with the entire class. Another option is to share the Game with a few children who then teach the Game to others. "Pair players" refers to players who collaborate to play against another pair. This recommended arrangement promotes mathematical thinking and communication as children develop and share successful strategies. "Pairs" refers to two children who collaborate to play the game and do not compete against each other. Some Games include easy versions as well as more challenging versions. Most Games require approximately 20 minutes of playing time. Games are ideal for home use since they provide children with additional practice and reassure parents that the number strand continues to be valued. When sending gameboards home, be sure to include the directions.

Games and Sponges provide children with a powerful vehicle for assessing their own mathematical abilities. During the Games, children receive immediate feedback that allows them to revise and to correct inefficient and inadequate practices. Appealing and repeatable, these Sponges and Games are valuable as center activities. Sponges and Games differ from the Independent Activities since they usually need to be introduced by a leader.

Independent Activities

Independent Activity sheets provide facts and computation practice. These sheets are designed to encourage practice of many more facts than would seem apparent at first glance. Some Independent Activity sheets allow multiple solutions. Most children will complete an Independent Activity sheet in 15 to 20 minutes. Independent Activity sheets can be completed in class or sent home as homework. Many Independent Activities provide two versions to accommodate different levels of difficulty and can be easily modified to provide additional practice.

Matrix of Games and Sponges

Type	Title	Topic	Page	Materials	Class	Pairs
S	**50 Chart Pieces**	Number Relationships	106	50 Chart, Counters, Form p. 107	✓	
S	**Action Counting**	Counting Forward and Backward	12	50 Chart, 3 × 5 Cards, Transparent Counters, Form p. 13	✓	
G	**Add or Subtract**	Addition and Subtraction Facts	92	Number Cubes*, Counters, Gameboards pp. 93-94		✓
S	**Box Sums**	Addition Facts	35	Dot Pattern Cards, Digit Cards, 5 × 8 Cards, Form p. 36	✓	
G	**Capture Two**	More and Less	26	Digit Cards		✓
G	**Choose and Subtract**	Subtraction Facts	69	Number Cubes*, Counters, Gameboards pp. 70-71		✓
G	**Choose Two**	Addition Facts	51	Number Cubes, Counters, Gameboard p. 52		✓
G	**Claim All You Can**	Number Relationships	115	Counters, Game Markers, Paper Clips/Pencils, Gameboard p. 116	✓	
S	**Coin Bingo**	Coin Recognition	126	Coins, Counters, Prepared Clues, Form p. 17	✓	
S	**Coin Count**	Adding Like Coins	127	50 Chart, Overhead Coins	✓	
G	**Coin Draw**	Adding Coins	132	Coins, Counters, Gameboard p. 135		✓
G	**Count and Cover 30**	Counting Forward and Backward	24	Number Cubes*, Counters, Gameboard p. 25		✓
G	**Cover Ten**	Addition and Subtraction Facts	90	Dot Cubes, Counters, Gameboard p. 91		✓
S	**Disappearing Robot**	Subtraction Facts	61	Chalkboard, Chalk, Chalk Eraser	✓	
S	**Find My Number**	Number Relationships	108	50 Chart, Counters, Paper Clip/Pencil, Form p. 109	✓	
G	**Finding Addends**	Addition Facts	49	Digit Cards, Counters, Gameboard p. 50		✓
G	**Five Plus**	Addition Facts	42	Ten Frame, Dot Cube*, Counters, Gameboard p. 43		✓
G	**How Many More?**	Subtraction Facts	72	Number Cubes, Counters, Gameboards pp. 73-74		✓
S	**How Many Now?**	Addition and Subtraction Facts	82	Ten Frame, Counters	✓	
G	**Less or More Spin**	More and Less	21	Number Cubes*, Counters, Paper Clips/Pencils, Gameboards pp. 22-23	✓	
G	**Low-High Spin**	Place Value	117	100 Chart, Paper Clips/Pencils, Paper Squares, Gameboard p. 118	✓	
G	**Make Ten**	Addition Facts	44	Dot Cubes, Counters, Gameboard p. 45		✓
S	**Name the Amount**	Number Sense to 100	102	100 Chart, Counters, 5 × 8 Cards, Form p. 103	✓	
S	**Narrow the Range**	More and Less	15	50 Chart, 5 × 8 Cards, Paper Strips	✓	
G	**Pair Search**	Addition and Subtraction Facts	95	Number Cubes, Counters, Gameboard p. 96		✓
S	**Pocket Full of Coins**	Adding Like Coins	128	Overhead Coins, Form p. 129	✓	

* denotes special number or dot cube

G = Games S = Sponges

Matrix of Games and Sponges (continued)

G = Games S = Sponges

Type	Title	Topic	Page	Materials	Class	Pairs
G	**Race to 20¢**	Exchanging and Adding Coins	130	Number Cubes*, Counters, Coins, Gameboard p. 133		✓
G	**Race to 40**	Place Value	113	Mini Ten Frames, Number Cubes, Beans, Gameboard p. 114		✓
G	**Roll Ten**	Addition Facts	46	Number Cubes*, Pencils, Forms pp. 47-48		✓
S	**Seeking Differences**	Subtraction Facts	62	Dot Pattern Cards, Counters, Form p. 63	✓	
S	**Seeking Sums**	Addition Facts	37	Dot Pattern Cards, Counters, Form p. 38	✓	
S	**Show Me**	Place Value	104	Mini Ten Frames, 50 Chart, Beans, Counters Form p. 105	✓	
S	**Show Then Change**	Comparing Numbers	14	Ten Frame, Counting Cards, Counters	✓	
G	**Taking from Ten**	Subtraction Facts	67	Ten Frames, Dot Cubes*, Counters, Gameboard p. 68		✓
S	**Ten Frame Differences**	Subtraction Facts	60	Ten Frame, Counters, Number Cube	✓	
S	**Ten Frame Sums**	Addition Facts	34	Ten Frame, Number Cube, Counters	✓	
S	**Three-in-a-Row**	More and Less	16	Dot Pattern Squares, Prepared Clues, Form p. 17	✓	
S	**What Works?**	Addition and Subtraction Facts	85	Dot Pattern Cards, Form p. 86	✓	
S	**What's Your Difference?**	Addition and Subtraction Facts	83	Digit Squares, 5 × 8 Cards, Form p. 84	✓	

* denotes special number or dot cube

Suggestions for Using *Nimble with Numbers*

Materials Tips

An effort has been made to minimize the materials needed. When appropriate, blackline masters are provided. The last section of the book contains generic types of blackline masters, including patterns for number and dot cubes. The six-sectioned spinners (p. 148) can substitute for a number cube or die. The blank spinner can be used for the specially marked number cubes (5–10 or 1-1-2-2-3-3). A simple spinner, like the one shown, can be assembled using one of the blackline master spinner bases, a paper clip, and a pencil. Many spinners are included on gameboards and require a paper clip and pencil. To reduce the noise and confine the area where cubes are rolled, use a box with felt glued to its bottom or lid.

A few activities use the Digit Squares (p. 143). The familiar sets of 0–9 number tiles substitute well for Digit Squares. If Digit Squares are not available, take time now to duplicate a set on card stock for each child.

A few activities require Dot Pattern Cards (p. 145). Dot Pattern Card sets should also be duplicated on card stock. Dot Pattern Cards are also needed for some class Sponges. Teachers should cut 2 sets of Dot Pattern Cards apart, place them in an appropriate container (paper sack, coffee can, or margarine tub), and store them in a handy place. As children gain confidence, Digit Cards (p. 142) may be substituted for Dot Pattern Cards. Children will have more success with the money activities if they have access to real or play coins or coin representations (p. 140).

Various materials work as markers on gameboards—different types of beans, multi-colored cubes, buttons, or transparent bingo chips (our preference due to the see-through feature). For some activities, children will need scratch paper and pencils. It is assumed that an overhead projector is available, but a chalkboard may be substituted.

Recommended Uses

The repeatable nature of these activities makes them ideal for continued use at home. Encouraging children to use these activities at home serves a dual purpose: parents are able to assist their children in gaining competence with the facts, and parents are reassured as they see the familiar basics practiced. To support your work in this area, we have included a parent letter and a list of helpful open-ended questions.

Besides being a source for more familiar homework, these activities offer a wide variety of classroom uses. The activities can be effectively used by substitute teachers as rainy-day options or for a change of pace. Many activities are short-term and require little or no preparation, making them ideal for soaking up spare moments at the end or beginning of a class period. They also work well as choices for center or menu activities. When children are absent from school, include these activities in independent work packets. You may package these activities in manila envelopes or self-closing transparent bags to facilitate frequent and easy checkout. To modify the activities and to accommodate the needs of your children, you may easily change the numbers, operations, and directions.

Getting the Most from These Activities

It is important to focus on increasing childrens' awareness of the mathematics being learned. To do this, pose open-ended questions that promote reflection, communication, and mathematical connections. For example, after using *Find My Numbers,* a colleague asked her second graders, "What patterns did you notice on the 50 Chart?" A child answered, "Look! When it says 10 more, all you have to do is go down one box, and 10 less means you go up one box." After playing *Cover Ten,* a teacher asked her children, "What amounts were easy to make?" One first grader answered, "Whatever I covered first was easy. When I had a few left, it was hard. Boy, that's a lot of adding and subtracting to find one answer." Children used a lot of trial and error to find answers.

Having children work together as pair players is of great value in increasing student confidence. While working this way, children have more opportunities to communicate strategies and to verbalize thinking. When asked to identify and to share their successful Game strategies verbally and in writing, children grow mathematically. Also, it is worthwhile to ask children to improve these activities or to create different versions of high-interest games.

Good questions help children make sense of mathematics, build their confidence, and encourage mathematical thinking and communication. A list of helpful, sample questions appears on page 10. Since the teacher's or parent's response impacts learning, we have included suggestions for responding. Share this list with parents for their use as they assist children with these activities and other unfamiliar homework tasks. This list was created by Leigh Childs for parent workshops and for inclusion in the California Mathematics Council's *They're Counting on Us, A Parent's Guide to Mathematics Education.* We have adapted the list for use with this book.

Parent Support

Parent Involvement

Since most parents place a high priority on attention to the number strand, they will appreciate the inviting and repeatable activities in this book. Because most parents willingly share the responsibility for repeated, short periods of practice, the *Family Letter* (p. 9) and the *Questions Sampler* (p. 10) were designed to promote parent involvement. The first home packet might include the *Family Letter,* the *Questions Sampler,* and *Less or More Spin* (pp. 21–23). To facilitate use, children should take home materials for making two Number Cubes (1–6 and 4–9). Since many children will benefit by practicing addition and subtraction facts, the next home packet might include *Finding Addends* (pp. 49–50) and a set of Digit Cards. Advise children and their families to keep the Number Cubes and Digit Cards in a safe place for frequent use throughout the school year.

Children enjoy and benefit from repeated use of *Seeking Sums* (pp. 37–38) and *What Works?*(pp. 85–86). These Sponges lend themselves to home packets as well. The advantage of Sponges, unlike Games, is that many can be experienced while a monitoring family member prepares dinner, packs lunches, or attends to other household tasks.

Concluding Thought

We hope that by using these materials, your children will develop more positive feelings towards mathematics as they improve their confidence and number competence.

Family Letter

Dear Family,

To be prepared to work in the 21st century, all children need to be confident and competent in mathematics. Today the working world requires understanding of all areas of mathematics including statistics, logic, geometry, and probability. To be successful in these areas, children must know their basic facts and be able to compute. It is important that we be more efficient and effective in the time we devote to arithmetic. You can help your child in this area.

Throughout the school year, our mathematics program will focus on enhancing your child's understanding of number concepts. However, children must devote time at school and at home to practice and to improve these skills. Periodically, I will send home activities and related worksheets that will build number sense and provide much needed practice. These games and activities have been carefully selected to engage your child in practicing more math facts than usually answered on a typical page of drill or during a flash-card session.

By using the enclosed *Questions Sampler* during homework sessions, you will be able to assist your child without revealing the answers. The questions are categorized to help you select the most appropriate questions for your situation. If your child is having difficulty getting started with a homework assignment, try one of the questions in the first section. If your child gets stuck while completing a task, ask one of the questions from the second section. Try asking one of the questions from the third section to have your child clarify his or her mathematical thinking.

Good questions will help your child make sense of the mathematics, build confidence, and improve mathematical thinking and communication. I recommend posting the *Questions Sampler* in a convenient place, so that you can refer to it often while helping your child with homework.

Your participation in this crucial area is most welcome.

Sincerely,

Questions Sampler

Getting Started

How might you begin?

What do you know now?

What do you need to find out?

While Working

How can you organize your information?

How can you make a drawing (model) to explain your thinking?

What approach (strategy) are you developing to solve this?

What other possibilities are there?

What would happen if . . . ?

What do you need to do next?

What assumptions are you making?

What patterns do you see? . . . What relationships?

What prediction can you make?

Why did you . . . ?

Checking Your Solutions

How did you arrive at your answer?

Why do you think your solution is reasonable?

What did you try that didn't work?

How can you convince me your solution makes sense?

Expanding the Response

(To help clarify your child's thinking, avoid stopping when you hear the "right" answer and avoid correcting the "wrong" answer. Instead, respond with one of the following.)

Why do you think that?

Tell me more.

In what other way might you do that? What other possibilities are there?

How can you convince me?

Counting and Comparing

Assumptions Children have been given many opportunities to count concrete objects, emphasizing understanding. Children have counted aloud in a variety of ways—by ones, forward and backward, and by using skip counting patterns. Children have also compared quantities and numbers and identified more and less. Visual models, such as 50 charts or 100 charts and number lines, have been used extensively.

Section Overview and Suggestions

Sponges

Action Counting pp. 12–13

Show Then Change p. 14

Narrow the Range p. 15

Three-in-a-Row pp. 16–17

These open-ended, whole-class, or small-group warm-ups actively engage children in practicing counting and comparing numbers. The frequent use of these Sponges will ensure greater success with the Games and Independent Activities in this section.

Skill Checks

Count On 1–6 pp. 18–20

These provide a way to help parents, children, and you to see children's improvement with counting and comparing numbers. Remember to have children respond to the STOP, the number sense task, before they solve the ten problems.

Games

Less or More Spin pp. 21–23

Count and Cover 30 pp. 24–25

Capture Two p. 26

These open-ended and repeatable Games actively involve children in counting and comparing numbers. All Games include tips that adapt the playing difficulty. *Less or More Spin* provides an additional challenging gameboard.

Independent Activities

More or Less pp. 27–28

Before, After, Between pp. 29–30

Which Numbers Fit? pp. 31–32

These Independent Activities provide long-term practice with counting and comparing numbers. *More or Less* and *Before, After, Between* are open-ended and reusable; in these activities, children create their own problems by rolling a number cube. *Which Number Fits?* provides practice with a number line. Each activity has two worksheets to vary the level of difficulty.

Action Counting

Topic: Counting Forward and Backward

Object: Count to and from numbers to 30.

Groups: Whole class or small group

__Tips__ Children will enjoy making up their own motions for future counting. Leader should not expect good physical coordination with all children; accept what he or she can do. Increase the difficulty by extending the counting range beyond 50 or 100.

Materials

- transparency of *Action Counting* form, p. 13
- three 3-by-5-inch cards to cover action choices on the form
- 2 transparent counters
- 50 Chart, p. 138

Directions

1. The leader displays one of the four action choices and leads children in practicing the action.
2. The leader selects and records a counting range to indicate the numbers that children will count to and from. A range of eight to fifteen counting numbers works well.
3. The leader directs children to count forward from the lowest number to the highest number with the action motions and then to count backward from the highest number to the lowest number with a reverse or related action.

   Example: Children count by ones in unison from 18 to 27. As they say each number, they cross stretch up in the air, alternating from right to left. When children reach 27, they count backwards from 27 to 18 and cross stretch to their toes as they say each number, alternating from right to left.

4. The leader continues to follow this procedure of displaying one action choice, recording a counting range, and leading children in counting forward and backward with the action motions.
5. When children are successful with this activity, the leader can indicate the counting range by displaying a 50 Chart and placing a transparent marker on each of two numbers. As children are led by a classmate to count with the action motions, another child can point to each corresponding number on the 50 Chart.

Making Connections

Promote reflection and make mathematical connections by asking:

- Which numbers were harder to remember? Why?
- How was the 50 Chart helpful?

Action Counting

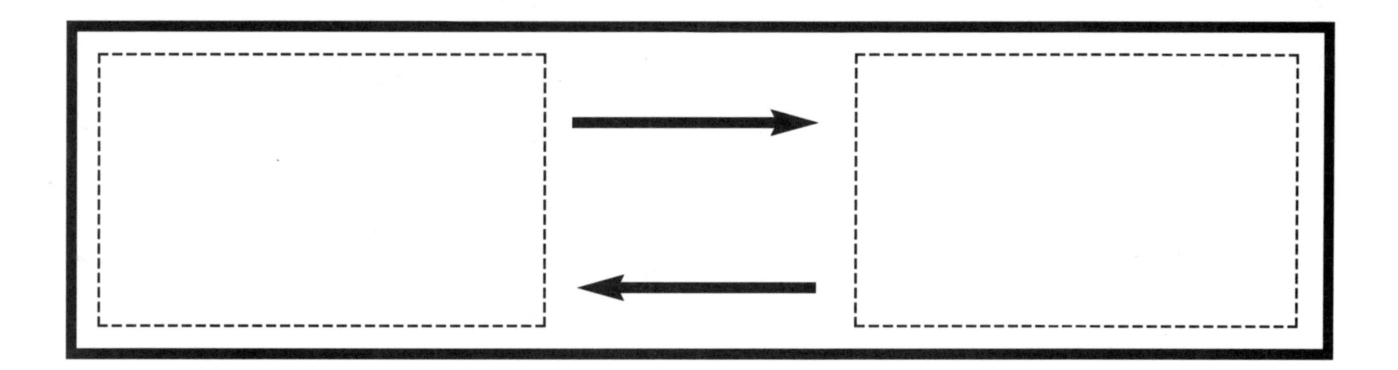

Front and Back Clap

Forward: Clap hands in front of body.
Backward: Clap hands behind body.

Cross Stretch

Forward: Reach high, alternating arms.
Backward: Reach low, alternating arms.

Arm Circles

"13,14,15"

"15,14,13"

Forward: Circle arms in forward direction.
Backward: Circle arms in backward direction.

March Tall, March Low

Forward: March in place, standing tall.
Backward: March in place, squatting low.

Show Then Change

Topic: Comparing Numbers

Object: Change an amount to a lesser or greater amount.

Groups: Whole class or small group

Materials

- transparency of Ten Frame form, p. 149
- Ten Frame form for each child, p. 149
- ten counters for each child
- Counting Cards 1–10, p. 141

Directions

1. The leader mixes and stacks the Counting Cards, draws one card, and announces the number. The leader places that many counters on the Ten Frame, filling in left to right and top row to bottom row.
2. The leader draws another Counting Card and announces the number.
3. The leader directs children to look at the Ten Frame and decide if the new amount is more or less than the original amount. Children respond, "More" or "Less," and the leader changes the Ten Frame to the new amount.

   Example: 7 counters are displayed on the Ten Frame. If the leader draws and announces "3," children announce "Less." The leader removes 4 counters to display 3 counters.

4. When the leader has modeled showing and changing amounts, and how to place counters on the Ten Frame, each child uses a Ten Frame and ten counters. The leader mixes and stacks the Counting Cards, draws one card, and tells children to show that many counters on their Ten Frame.
5. The leader continues drawing a Counting Card and announcing the number as children decide if they make the new amount by placing more or less counters, respond "More" or "Less," and then change their Ten Frames to the new amount.
6. When children are successful with this Sponge, larger numbers can be practiced with Counting Cards 1–20 and two Ten Frames.

Tip *This sponge can be adapted into a game by having children work in pairs. One child draws a counting card, while the other child shows the amount on the Ten Frame with counters. On the next draw, children reverse rolls, with the second child drawing and the first child announcing "More" or "Less" and changing counters to the new amount.*

Making Connections

Promote reflection and make mathematical connections by asking:

- What helps you determine whether the new amount is more or less?

Narrow the Range

Topic: More and Less

Object: Narrow the range of numbers to identify one number.

Groups: Whole class or small group

Materials

- transparency of 50 Chart, p. 138
- two 5-by-8-inch cards
- two 1-by-8-inch paper strips

Tips *After children are familiar with this activity, give each child a 50 Chart and cards and strips to mask numbers they eliminate. Increase the difficulty by using a 100 Chart to extend the counting range beyond 50.*

Directions

1. The leader displays the 50 Chart, announcing "I am thinking of a number. My number is less than 21. What numbers can we eliminate on this chart? Let's narrow the range." After discussion, the leader covers the bottom three rows with a card, hiding the numbers 21 through 50 and leaving 1 through 20 displayed.
2. The leader encourages children to ask more-or-less questions, such as "Is it more than ____?" or "Is it less than ____?" The leader responds, "Yes, my number is more/less than ____." or "No, my number is not more/less than ____."

   Example: The numbers 1–20 are displayed; the secret number is 17. If a child asks, "Is it more than 5?" the leader would respond, "Yes, it is more than 5."

					6	7	8	9	10
11	12	13	14	15	16	17	18	19	20

3. The leader announces, "Let's narrow the range. What numbers can we eliminate?" After discussion, the leader hides the numbers 1 through 5 with a paper strip, leaving 6 through 20 displayed.
4. Children continue asking more-or-less questions, listening to the leader's answer, and discussing which numbers to eliminate. Each guess should narrow the range.
5. Play continues until a child guesses the correct number.
6. The child with the correct guess whispers a new secret number to the leader, and the leader begins a new round of play.

Making Connections

Promote reflection and make mathematical connections by asking:

- If a number is less than 10, can 10 be the number? Why or why not?
- How was the 50 Chart helpful?

Three-in-a-Row

Topic: More and Less

Object: Identify one and two more or one and two less.

Groups: Whole class or small group

Materials

- prepared clues
- *Three-in-a-Row* form for each child, p. 17
- Dot Pattern Squares for each child, p. 146
- container for 1–9 Dot Pattern Squares

Directions

1. Children randomly arrange their nine Dot Pattern Squares on *Three-in-a-Row* forms to create unique playing boards.
2. The leader draws a Dot Pattern Square and reads the clue.

   Clues
   1 one less than the number of socks in a pair
   2 one more than the number of noses on a face
   3 two less than the number of toes on a foot
   4 two more than the number of eyebrows on a face
   5 two more than the number of wheels on a tricycle
   6 one less than the number of days in a week
   7 one less than the number of legs on two chairs
   8 two less than the number of pennies equal to a dime
   9 one more than the number of legs on a spider
3. Children individually try to identify the amount and turn that card facedown on their playing boards. Children agree on the correct response to the first clue before leader reads a second.
4. The leader continues giving clues as children locate and turn the corresponding cards to the blank sides.
5. When children have three blank cards in a row, they call out, "Three-in-a-row." The game continues until all children have at least one "three-in-a-row."
6. Children enjoy repeated rounds of this Sponge as the activity often concludes after six or seven clues are shared.

Tips *Activity works faster and well with simple numerical clues like "two more than six" and "one less than four." First graders could begin with all clues being "one more than ___." Children enjoy collaborating to create clues for future warm-ups, and leaders can model with personalized clues for their classes ("two more than the number of hamsters in our cage").*

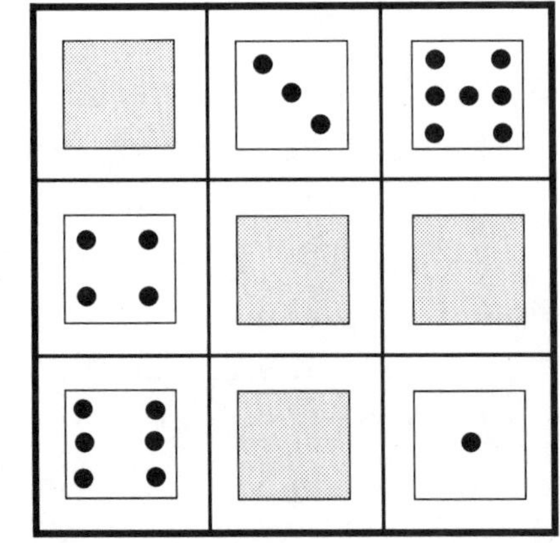

Making Connections

Promote reflection and make mathematical connections by asking:

- What did you do mentally to figure out the clue?
- How could you keep track of more than one three-in-a-row?

Three-in-a-Row

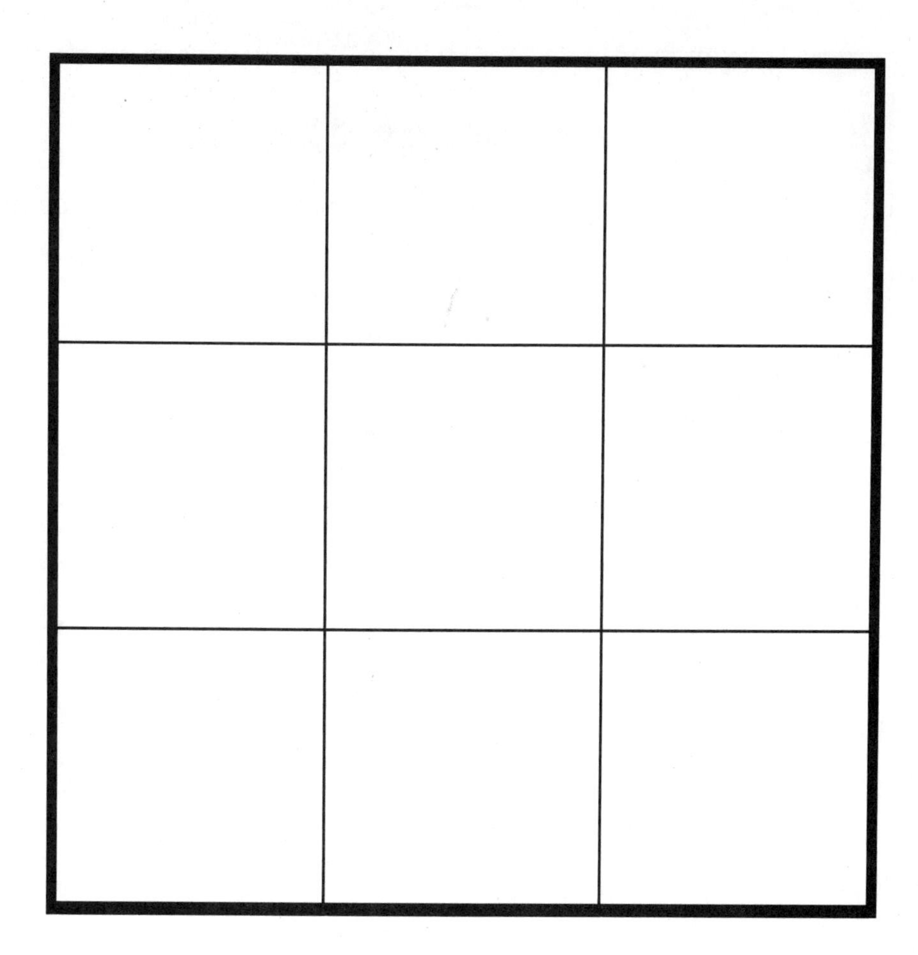

Three-in-a-Row

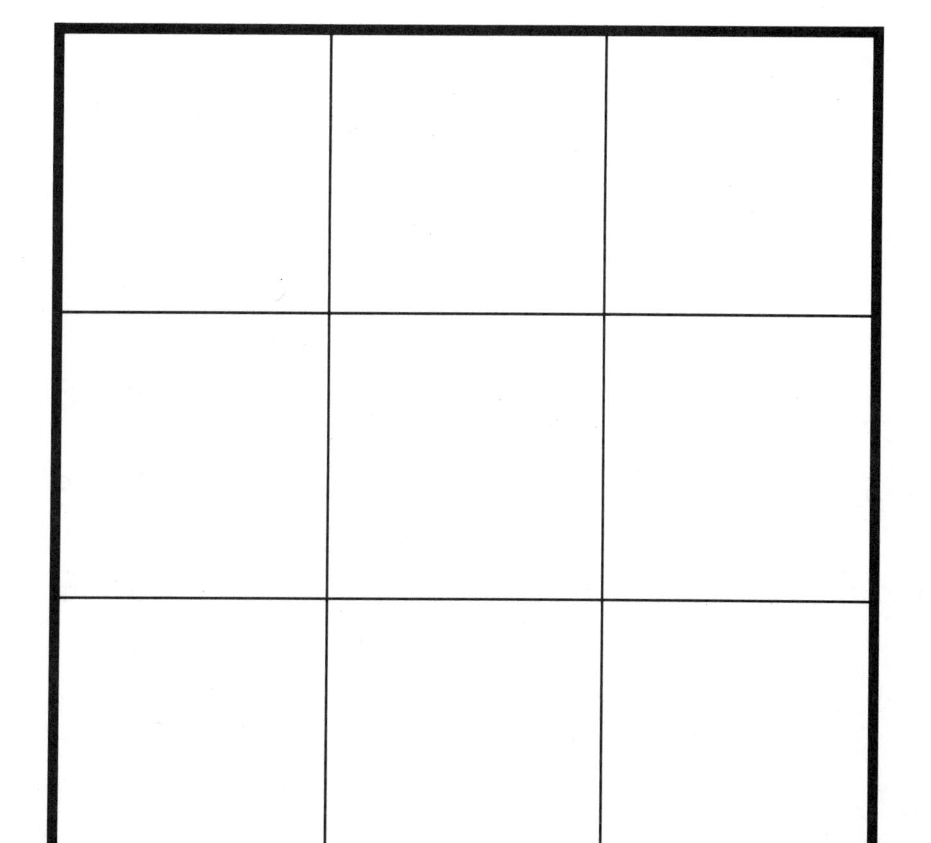

Date ____________ Name ____________

Count On 1

STOP Don't start yet! Circle a problem that may have an answer more than 10.

Write the missing numbers.

1. 5 ____ ____ 8
2. 6 ____ 4
3. 16 17 ____
4. 18 17 16 ____
5. 1 more than 5 ____
6. 1 less than 9 ____
7. 2 more than 13 ____
8. 2 less than 14 ____

Circle numbers 2 apart.

9. 2 4 7
10. 11 13 16

Go On What numbers are missing? 16, ____, 14, ____, 12, ____, 10, ____

Date ____________ Name ____________

Count On 2

STOP Don't start yet! Circle a problem with a counting sequence that may get smaller.

Write the missing numbers.

1. 1 ____ 3 ____
2. 7 ____ 5
3. 11 ____ 13
4. 15 14 ____ 12
5. 1 less than 8 ____
6. 2 more than 6 ____
7. 2 less than 13 ____
8. 1 more than 16 ____

Circle numbers 2 apart.

9. 5 6 8
10. 17 18 19

Go On Which numbers are 2 apart? | 4 7 8 9 | | 2 5 4 8 |

Date ____________ Name ____________________

Count On 3

STOP Don't start yet! Circle a problem with a counting sequence that may get larger.

Write the missing numbers.

1. 3 _____ _____ 6

2. 9 _____ 7

3. 14 15 _____

4. 14 13 12 _____

5. 2 more than 7 _____

6. 2 less than 8 _____

7. 1 more than 12 _____

8. 1 less than 16 _____

Circle numbers 2 apart.

9. 4 5 6

10. 12 13 15

Go On What numbers are missing? _____, 17, _____, 15, _____, 13, _____, 11

Date ____________ Name ____________________

Count On 4

STOP Don't start yet! Circle a problem with a counting sequence that may get smaller.

Write the missing numbers.

1. 6 7 _____ _____

2. 5 _____ 3

3. 18 _____ 20

4. 20 19 _____ 17

5. 2 less than 10 _____

6. 1 more than 9 _____

7. 1 less than 17 _____

8. 2 more than 16 _____

Circle numbers 2 apart.

9. 2 5 7

10. 14 16 17

Go On Look for a pattern.
Write 2 more number pairs that belong.

9, 11	6, 8	14, 16

Date ____________ Name ____________________

Count On 5

STOP Don't start yet! Circle a problem that may have an answer less than 10.

Write the missing numbers.

1. 7 ____ ____ 10
2. 10 ____ 8
3. 12 ____ 14
4. 17 16 15 ____
5. 1 more than 6 ____
6. 2 less than 10 ____
7. 2 more than 17 ____
8. 1 less than 11 ____

Circle numbers 2 apart.

9. 3 4 5
10. 12 14 17

Go On What numbers are missing? 19, 18, ____, ____, 15, ____, ____, 12

Date ____________ Name ____________________

Count On 6

STOP Don't start yet! Circle a problem with a counting sequence that may get smaller.

Write the missing numbers.

1. 4 ____ 6 ____
2. 8 ____ 6
3. 16 17 ____
4. 13 12 ____ 10
5. 1 less than 6 ____
6. 2 more than 8 ____
7. 2 less than 18 ____
8. 1 more than 16 ____

Circle numbers 2 apart.

9. 4 7 9
10. 16 18 19

Go On Look for a pattern.
Write 2 more number pairs that belong.

7, 6	18, 17	12, 11

Less or More Spin

Topic: More and Less

Object: Cover three numbers in a row with your counters.

Groups: 2 pair players

Tip *If the game is played at home, require adult players to get four-in-a-row.*

Materials for each group

- *Less or More Spin A* gameboard, p. 22
- Number Cube (1–6), p. 147
- counters (different kind for each pair)
- pencil and paper clip for spinner

Directions

1. The first pair rolls the Number Cube and spins the spinner. The Number Cube indicates the starting number and the spinner indicates the direction. By using the rolled number and the spun direction, the pair identifies possible numbers to cover.

   Example: If "3" is rolled and "1 less" is spun, the pair locates the 2s on the gameboard. However, if "3" is rolled and "less" is spun, the pair locates 0s, 1s, and 2s.

2. The first pair discusses and decides which one number to cover with a counter and then covers that number.
3. The second pair rolls and spins to determine possible numbers they can cover. After discussion, the pair selects and covers one number.
4. Pairs continue to alternate turns, rolling, spinning, and determining where to place the counter.
5. If a rolled and spun combination does not result in any available number to cover, the pair rolls again.
6. The first pair to have three counters in a row, horizontally, vertically, or diagonally, wins.
7. The *Less or More Spin B* gameboard, p. 23, is a more challenging version since children use a 4–9 Number Cube and a spinner with additional options.

7	3	6	0
4	5	1	2
3	7	6	5
1	0	2	4

Making Connections

Promote reflection and make mathematical connections by asking:

- What strategies helped you line up your counters in a row?
- How did you select which number to cover with your counter?

Less or More Spin A

7	3	6	0
4	5	1	2
3	7	6	5
1	0	2	4

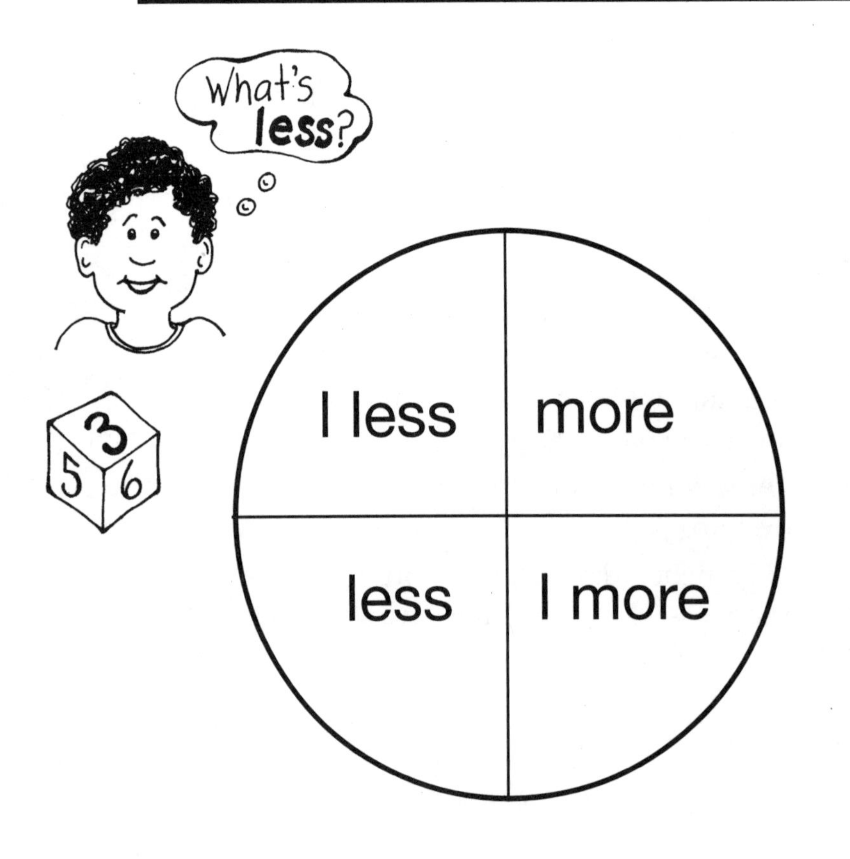

Less or More Spin B

0	3	4	10	6
13	6	8	1	7
10	2	9	5	12
8	4	7	6	3
7	14	11	5	9

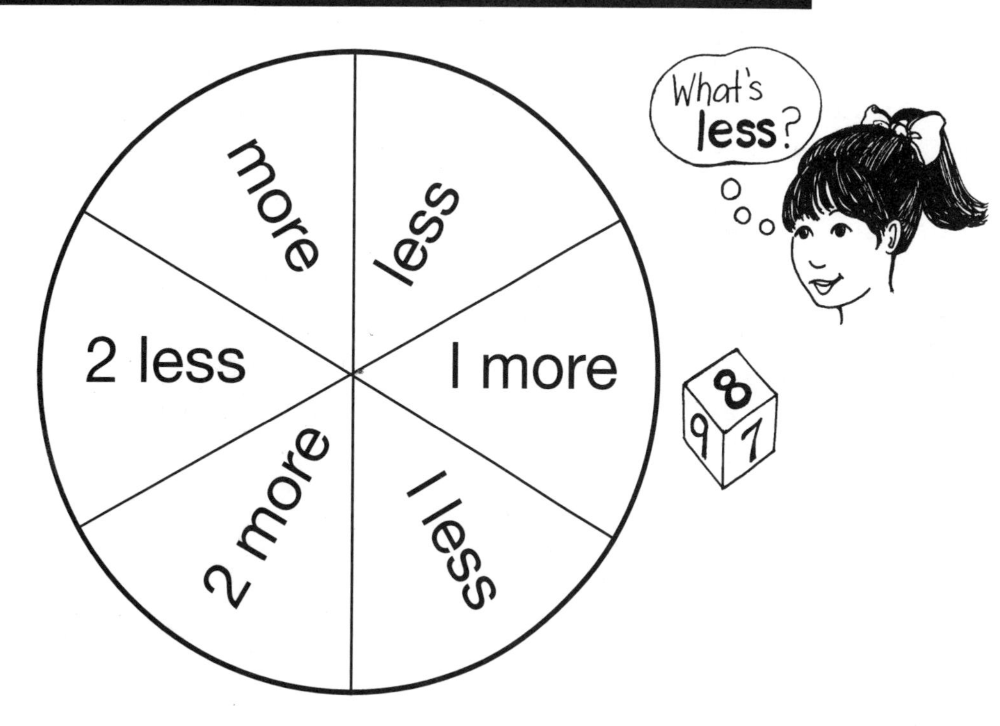

Count and Cover 30

Topic: Counting Forward and Backward

Object: Count and cover 30 cells in sequence.

Groups: Pairs

Materials for each pair

- *Count and Cover 30* gameboard, p. 25
- special Number Cube (1-1-2-2-3-3), p. 147
- 30 counters

Count and Cover 30
1. Roll
2. Predict
3. Take counters
4. Announce amount on the board
5. Count on

Directions

1. In this game, a pair works cooperatively to count to 30.
2. One child rolls the Number Cube to determine how many cells to cover and points to where the last counter might be placed.
3. This child takes the indicated number of counters and counts by ones while individually placing the counters on the gameboard.
4. The other child rolls the Number Cube and predicts where the last counter will be placed. The child takes the indicated number of counters, announces the amount already placed, and counts on by ones as counters are individually placed on the gameboard.

   Example: 2 counters are on the gameboard. The second child rolls a 3. The child points to the 5-cell to predict where he/she expects to land, then takes 3 counters, announces "2," and counts on "3, 4, 5" while placing counters on the gameboard.
5. Members of the pair alternate turns as they roll, predict, and count on and up to 30.
6. If a child seems confused counting on or placing counters, the members of the pair may collaborate to determine the correct sequence, and/or refer to the chart shown above.
7. The child who counts to 30 first may play first in the next round.
8. To help children count backward, begin a round with a counter on "30" and use a rolled number to determine how many cells to cover from 30 to 1. The child follows the same procedure for predicting and placing counters.

Tip *To make this game accommodate more experienced players, extend the counting sequence beyond 30 by using a 50 Chart or 100 Chart.*

Making Connections

Promote reflection and make mathematical connections by asking:

- What helped you correctly identify the starting number?
- What helped you predict the placement of the last counter?

Count and Cover 30

				5					10
				15					20
				25					30

Capture Two

Topic: More and Less

Object: Capture number pairs that are two apart.

Groups: 2 players

Materials for each group

- 2 sets of Digit Cards (0–9), p. 142

Directions

1. Randomly mix and stack Digit Cards facedown.
2. The first player draws and displays two Digit Cards. If the digits on the cards are two apart, the player captures the cards by stating their relationship. If the two Digit Cards are more or less than two apart, the cards remain displayed for future plays.

   Example: If 6 and 4 are displayed, the player says "4 is 2 less than 6" (or "6 is 2 more than 4") and takes and keeps both cards.
3. The second player draws and displays one or two Digit Cards. (The second player will draw one card from the stack if any cards remain displayed from the previous play. The same player will draw two cards if no cards remain displayed from the previous play.)
4. If a drawn card can be paired with one of the displayed cards, the relationship is stated and the two cards are kept by the player.
5. Players continue to alternate turns, drawing and displaying cards, and stating paired relationships when possible.
6. After the stack of Digit Cards is depleted, the players count their captured cards to see who has the most.

Tip *To make this game accommodate less-experienced players, have children pair cards that are only one apart and/or replace Digit Cards with Dot Pattern Cards, p. 145.*

Making Connections

Promote reflection and make mathematical connections by asking:

- What approach helped you find pairs?

Date ____________ Name ____________

More or Less I

Roll a number cube. Write your rolled number in a box. Read the directions and write the correct number on the line. Will the number 9 ever be an answer?

Write 1 more.	Write 1 less.
☐ ____	☐ ____
☐ ____	☐ ____
☐ ____	☐ ____
☐ ____	☐ ____
Write 2 more.	**Write 2 less.**
☐ ____	☐ ____
☐ ____	☐ ____
☐ ____	☐ ____
☐ ____	☐ ____

Date ____________ Name ____________________

More or Less II

Roll a number cube. Write your rolled number in a box. Read the directions and write the correct number on the line. Will zero ever be an answer?

1 more than ☐ _____	1 less than ☐ _____
2 more than ☐ _____	2 less than ☐ _____
1 more than ☐ _____	1 less than ☐ _____
2 more than ☐ _____	2 less than ☐ _____
1 more than ☐ _____	1 less than ☐ _____
2 more than ☐ _____	2 less than ☐ _____
1 more than ☐ _____	1 less than ☐ _____
2 more than ☐ _____	2 less than ☐ _____
1 more than ☐ _____	1 less than ☐ _____

Date ____________ Name ____________

Before, After, Between I

Roll a number cube. Write your rolled number in a box. On a line that follows the box, write the number that comes after. On a line before a box, write the number that comes before.

1. ☐ ____	2. ☐ ____
3. ☐ ____	4. ☐ ____
5. ☐ ____	6. ☐ ____
7. ____ ☐	8. ____ ☐
9. ____ ☐	10. ____ ☐
11. ____ ☐	12. ____ ☐
13. ____ ☐ ____	14. ____ ☐ ____
15. ____ ☐ ____	16. ____ ☐ ____
17. ____ ☐ ____	18. ____ ☐ ____

Independent Activity

Date ____________ Name ____________

Before, After, Between II

Roll a number cube. Write your rolled number in a box. On a line that follows the box, write the number that comes after. On a line before a box, write the number that comes before.

1. ____ ☐	2. ☐ ____
3. ____ ☐ ____	4. ____ ☐
5. ☐ ____ ____	6. ☐ ____
7. ____ ☐ ____	8. ____ ☐
9. ☐ ____	10. ____ ☐ ____
11. ____ ____ ☐	12. ☐ ____ ____
13. ____ ☐	14. ☐ ____
15. ____ ☐ ____	16. ____ ☐
17. ☐ ____ ____	18. ____ ☐ ____

Independent Activity

Date ______________ Name ______________________

Which Numbers Fit? I

Circle the numbers that fit.

1.	More than 6 0 1 2 3 4 5 6 7 8 9 10
2.	Less than 4 0 1 2 3 4 5 6 7 8 9 10
3.	More than 3 0 1 2 3 4 5 6 7 8 9 10
4.	Less than 8 0 1 2 3 4 5 6 7 8 9 10
5.	More than 7 0 1 2 3 4 5 6 7 8 9 10
6.	Less than 9 0 1 2 3 4 5 6 7 8 9 10
7.	More than 5 0 1 2 3 4 5 6 7 8 9 10
8.	Less than 10 0 1 2 3 4 5 6 7 8 9 10
9.	More than 0 0 1 2 3 4 5 6 7 8 9 10
10.	Less than 2 0 1 2 3 4 5 6 7 8 9 10

Date ____________ Name ____________________

Which Numbers Fit? II

Circle the numbers that fit.

1. More than 13

10 11 12 13 14 15 16 17 18 19 20

2. Less than 15

10 11 12 13 14 15 16 17 18 19 20

3. More than 16

10 11 12 13 14 15 16 17 18 19 20

4. Less than 17

10 11 12 13 14 15 16 17 18 19 20

5. More than 11

10 11 12 13 14 15 16 17 18 19 20

6. Less than 20

10 11 12 13 14 15 16 17 18 19 20

7. More than 14

10 11 12 13 14 15 16 17 18 19 20

8. Less than 12

10 11 12 13 14 15 16 17 18 19 20

9. More than 13

10 11 12 13 14 15 16 17 18 19 20

10. Less than 18

10 11 12 13 14 15 16 17 18 19 20

Addition Facts

Assumptions The addition facts have previously been taught and reviewed, with an emphasis on understanding. Concrete objects and visual models such as Ten Frames and dominoes have been used extensively. Children have used manipulatives to discover different combinations that can be made for a specific number.

Section Overview and Suggestions

Sponges

Ten Frame Sums p. 34

Box Sums pp. 35–36

Seeking Sums pp. 37–38

These open-ended, whole-class, or small-group warm-ups actively engage children in practicing many addition facts. The frequent use of these Sponges will ensure greater success with the Games and Independent Activities in this section.

Skill Checks

Just the Facts 1–6 pp. 39–41

These provide a way to help parents, children, and you to see children's improvement with the addition facts. Copies of *Just the Facts* may be cut in half so that each Check may be used at a different time. Remember to have children respond to STOP, the number sense task, before they solve the ten problems.

Games

Five Plus pp. 42–43

Make Ten pp. 44–45

Roll Ten pp. 46–48

Finding Addends pp. 49–50

Choose Two pp. 51–52

These open-ended repeatable Games actively involve children in practicing many addition facts. *Five Plus* uses a Ten Frame to help children visualize amounts. The challenging Games *Make Ten* and *Roll Ten* encourage flexible thinking and require children to combine addends to equal ten. The *Roll Ten* open-ended recording sheet allows practice of the more difficult facts. *Finding Addends* and *Choose Two* promote identification of addends. All Games promote mental computation as children enhance their strategic thinking skills.

Independent Activities

Seeking Sums Practice pp. 53-54

Roll and Fill pp. 55-56

Neighbor Sums pp. 57-58

Each Independent Activity has two worksheets that allow children to independently practice the easier and/or more difficult addition facts. *Seeking Sums Practice* uses a helpful, visual dot patterns format. *Roll and Fill* reuses the familiar steps followed in the Game *Roll Ten*, and includes an open-ended recording sheet to allow practice of additional facts.

Ten Frame Sums

Tip Facilitate success of this warm-up by displaying the number of counters rolled near the Ten Frame.

Topic: Addition Facts

Object: Visualize, verbalize, and display sums.

Groups: Whole class or small group

Materials

- Ten Frame form for each child, p. 149
- counters for each child
- transparency of Ten Frame form, p. 149
- counters
- Number Cube (1–6), p. 147

Directions

1. The leader places up to four counters in the displayed Ten Frame and asks children, "How many?"
2. Children display the same number of counters in their individual Ten Frames.
3. The Number Cube is rolled by a child, who announces the amount.
4. Using the rolled amount, the leader asks children to visualize adding that many more counters to the Ten Frame and to visualize the new sum.

   Example: If four is displayed and a three is rolled, children visualize a Ten Frame with seven counters.
5. After children state the sum and combination, children display the sum on their Ten Frames.
6. After children clear their Ten Frames, steps 1 through 5 are repeated. Additional rounds of this activity will help children mentally develop a valuable Ten Frame reference.

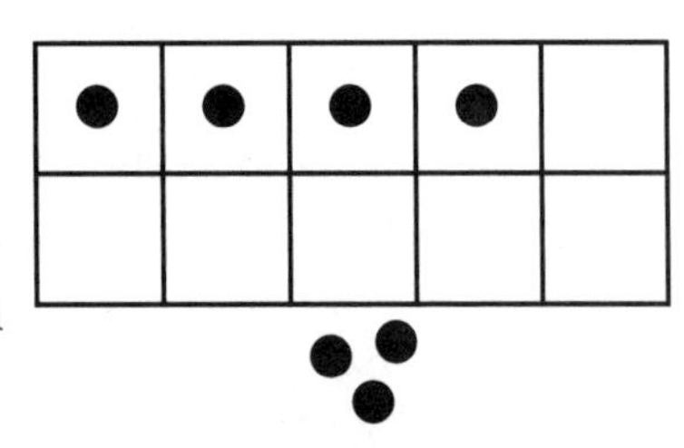

Making Connections

Promote reflection and make mathematical connections by asking:

- How did the Ten Frame help you identify the correct sum?
- Which amounts are easier to visualize? Why?

Box Sums

Topic: Addition Facts

Object: Mentally find sums in designated sections of a box square.

Groups: Whole class or small group

Materials

- transparency of *Box Sums*, p. 36
- transparency of Dot Pattern Cards, p. 145
- 5-by-8-inch card
- Digit Cards, p. 142 (set for each child)

Tip *When children seem ready, display Digit Cards in place of Dot Pattern Cards.*

Directions

1. The leader displays *Box Sums* activity sheet with Dot Pattern Cards placed in each of the four sections.
2. Using an index card, the leader hides the right half and displays the left half of the square and asks children, "What's the sum?" "What did you do mentally to find the sum?"
3. The leader continues this process for each half section. Children respond by using their Digit Cards to show the sums.
4. The leader displays the entire activity sheet and asks, "How many altogether?" "What did you do mentally to find the sum?"
5. The leader places new combinations of Dot Pattern Cards on the *Box Sums* activity form and displays different half sections. Children use their Digit Cards to show the sums.
6. The leader may repeat these steps with a variety of combinations.

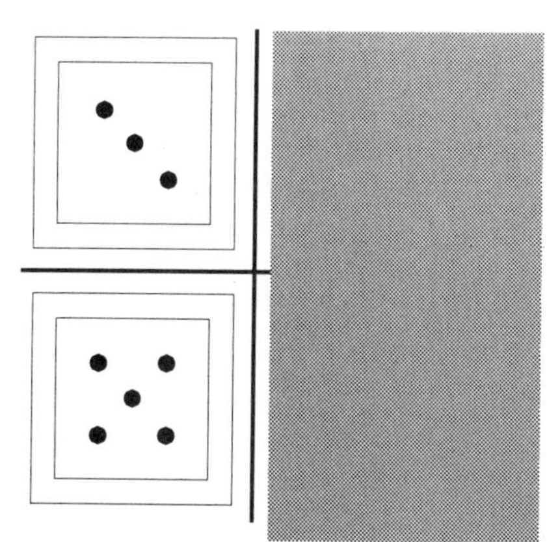

Making Connections

Promote reflection and make mathematical connections by asking:

- What strategies did you use to quickly identify sums?

Box Sums

Seeking Sums

Topic: Addition Facts

Object: Combine dots to equal target sums.

Groups: Whole class or small group

Materials

- transparent Dot Pattern Cards 1–6, p. 145
- transparency of *Seeking Sums*, p. 38
- 12 transparent counters

Tips To increase the difficulty, include a 7, 8, or 9 Dot Pattern Card, Transition into Digit Cards as children seem ready.

Directions

1. The leader randomly selects four Dot Pattern Cards and displays them on the *Seeking Sums* activity sheet.

2. Using any of the four displayed numbers, children identify ways to make the sums one through twelve.

Example: If 1, 2, 4, and 5 are displayed, 5 could be made two different ways (combining 1 and 4 or with the 5 card alone).

3. When a child identifies a way to show a sum, the leader covers the sum with a counter. After a solution is given, the leader should encourage players to seek additional solutions for the same sum.

Making Connections

Promote reflection and make mathematical connections by asking:

- Were any sums not possible? If so, why not?
- Which sums could be made more than one way?

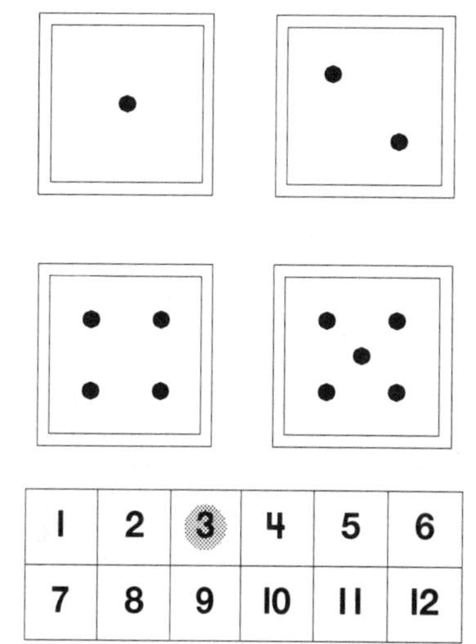

Seeking Sums

1	2	3	4	5	6
7	8	9	10	11	12

Date ____________ Name ____________________

Just the Facts 1

Don't start yet! Circle two problems that may have sums less than 10.

1. 3 + 2 = ___

2. 6 + 2 = ___

3. 5 + 5 = ___

4. 7 + 4 = ___

5. 3 + 3 = ______ **6.** 6 + 3 = ______ **7.** 4 + 3 + 2 = ______

8. 6 + 6 = ______ **9.** 4 + 5 + 2 = ______ **10.** 7 + 7 = ______

What numbers come next? 2, 4, 6, _____, _____, _____

What is the addition rule?

Date ____________ Name ____________________

Just the Facts 2

Don't start yet! Circle the problem in row one that may have the largest answer.

1. 2 + 4 = ___

2. 7 + 2 = ___

3. 6 + 3 = ___

4. 8 + 4 = ___

5. 5 + 3 = ______ **6.** 7 + 3 = ______ **7.** 2 + 3 + 5 = ______

8. 9 + 2 = ______ **9.** 6 + 2 + 4 = ______ **10.** 9 + 4 = ______

Go On: Write three addition facts that equal 8.

Date ________________ Name ______________________________

Just the Facts 3

Don't start yet! Circle two problems that may have sums greater than 10.

1. 5 + 2 = ___

2. 6 + 3 = ___

3. 5 + 7 = ___

4. 8 + 3 = ___

5. 3 + 4 = ______ **6.** 6 + 4 = ______ **7.** 3 + 3 + 3 = ______

8. 8 + 4 = ______ **9.** 3 + 3 + 5 = ______ **10.** 7 + 6 = ______

What numbers come next? 1, 3, 5, _____, _____, _____
What is the addition rule?

Date ________________ Name ______________________________

Just the Facts 4

Don't start yet! Circle two problems that may have even sums.

1. 3 + 5 = ___

2. 7 + 3 = ___

3. 6 + 6 = ___

4. 7 + 5 = ___

5. 4 + 5 = ______ **6.** 7 + 2 = ______ **7.** 3 + 3 + 4 = ______

8. 6 + 5 = ______ **9.** 3 + 4 + 5 = ______ **10.** 8 + 6 = ______

Write three addition facts that equal 9.

Date ____________ Name ____________

Just the Facts 5

Don't start yet! Circle two problems that may have odd sums.

1. $\begin{array}{r} 4 \\ +4 \\ \hline \end{array}$

2. $\begin{array}{r} 3 \\ +6 \\ \hline \end{array}$

3. $\begin{array}{r} 5 \\ +6 \\ \hline \end{array}$

4. $\begin{array}{r} 4 \\ +7 \\ \hline \end{array}$

5. 2 + 5 = ______

6. 4 + 6 = ______

7. 5 + 2 + 2 = ______

8. 9 + 3 = ______

9. 4 + 3 + 4 = ______

10. 8 + 5 = ______

What numbers come next? 1, 4, 7, ______, ______, ______

What is the addition rule?

Date ____________ Name ____________

Just the Facts 6

Don't start yet! Circle the problem in row one that may have the smallest answer.

1. $\begin{array}{r} 5 \\ +4 \\ \hline \end{array}$

2. $\begin{array}{r} 6 \\ +4 \\ \hline \end{array}$

3. $\begin{array}{r} 7 \\ +5 \\ \hline \end{array}$

4. $\begin{array}{r} 4 \\ +8 \\ \hline \end{array}$

5. 4 + 2 = ______

6. 3 + 7 = ______

7. 4 + 4 + 2 = ______

8. 8 + 3 = ______

9. 3 + 3 + 6 = ______

10. 9 + 5 = ______

Go On Write three addition facts that equal 10.

Five Plus

Topic: Addition Facts Beyond 5

Object: Fill a pathway to the star.

Groups: Pairs

Materials for each group

- *Five Plus* gameboard, p. 43
- Ten Frame, p. 149
- Dot cube (1-2-3-4-5-Choose), p. 144
- 30 counters (25 of one kind, 5 of a different kind)

Tips *As pairs play, one child might record the corresponding equations. To extend the playing time, have children attempt to complete two or more pathways.*

Directions

1. In this game, a pair works cooperatively to fill a pathway.
2. Place five counters of one kind in the top row of the Ten Frame. (The counters remain there during the entire game.)
3. The pair rolls the Dot Cube. If "Choose" appears, pair selects any number one through five. When a number is rolled, the pair adds that amount of counters to the Ten Frame, states the equation, and places a counter on the path with that sum.

   Example: If 4 is rolled, the Ten Frame is filled to show 9 (5 of one color, 4 of another), "Five plus four equals nine" is stated, and a counter is placed in the first cell of the 9-pathway.

4. The pair continues until one path to the star is filled.

Making Connections

Promote reflection and make mathematical connections by asking:

- How did the Ten Frame help you identify the correct sum and pathway?

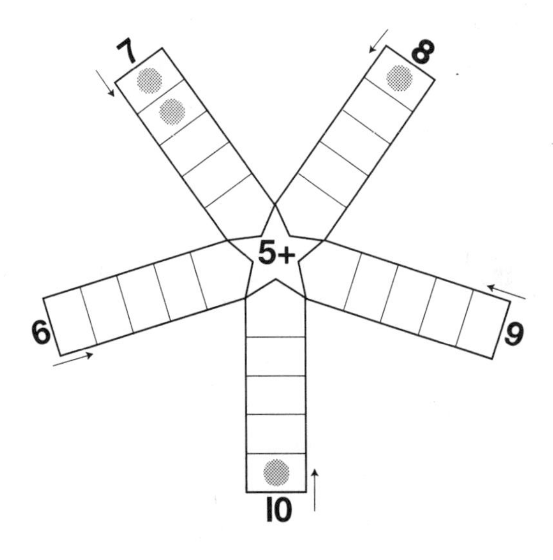

Five Plus

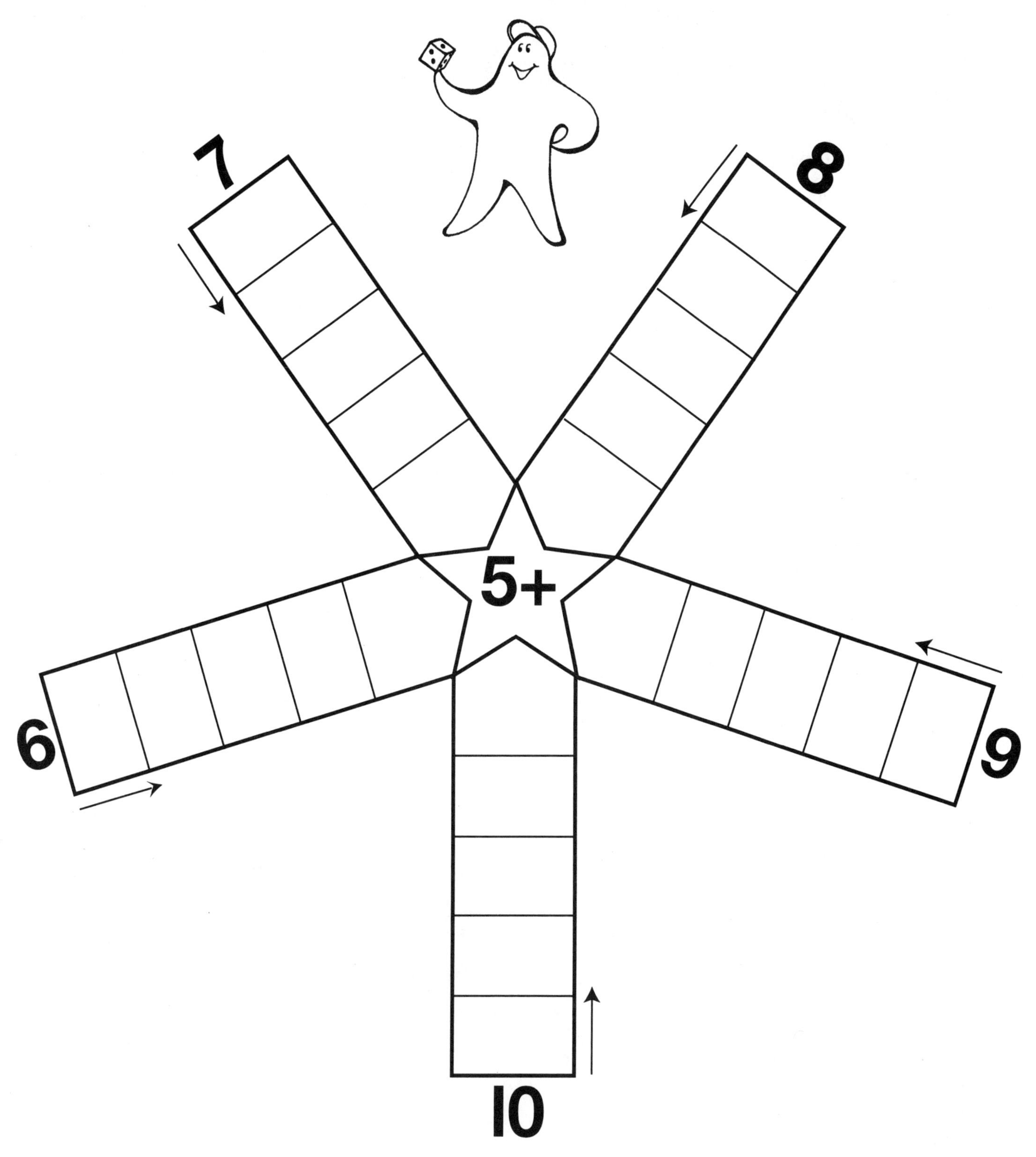

Make Ten

Topic: Addition Facts to Equal Ten

Object: Combine addends to cover all numbers.

Groups: Pairs

Materials for each group

- *Make Ten* gameboard, p. 45
- 2 Dot Cubes (1–6), p. 144
- counters

Tips *Replace one Dot Cube with a Number Cube. As children play, they might record their equations. For a challenge, have children roll three Dot Cubes, and use one, two, or all three numbers to see how many addends they can cover in one turn.*

Directions

1. In the game, a pair works cooperatively to cover all numbers.
2. The pair rolls two Dot Cubes.
3. The pair uses any number on the gameboard to add to one or two Dot Cubes to make ten. The pair states aloud the equation and covers the number with a counter.
4. The pair continues to use the same two amounts shown on the Dot Cubes and covers other numbers.

   Example: If 3 and 4 are shown on the Dot Cubes, children could cover 7 ($7 + 3 = 10$), 6 ($6 + 4 = 10$), and 3 ($3 + 3 + 4 = 10$).
5. When no other numbers can be covered, the pair rolls the two Dot Cubes and tries to cover additional numbers.
6. The pair continues to play until all numbers on the gameboard are covered.
7. The pair clears the gameboard and rolls two Dot Cubes to repeat the procedure.

9	8	7	6	5
4	3	2	1	0

Making Connections

Promote reflection and make mathematical connections by asking:

- With which rolls were you able to cover the fewest numbers? Why?

Make Ten

9	8	7	6	5
4	3	2	1	0

Make Ten

9	8	7	6	5
4	3	2	1	0

Roll Ten

Topic: Addition Facts

Object: Create equations to equal 10.

Groups: 2 players

Materials for each group

- *Roll Ten A* recording sheet for each child, p. 47
- Number Cube (3-4-5-6-7-Choose), p. 147
- pencil

Tip *Use "Roll ___," the open-ended recording sheet, to practice sums other than 10 and adjust the number cube accordingly.*

Directions

1. The first player rolls the Number Cube and records the resulting number as one of the addends.
2. The second player rolls the Number Cube and records similarly.
3. The first player rolls again. If the number rolled can be combined with the previously rolled number to equal 10, the player completes one of the four equations. If not, the player records the rolled number as an addend of another equation.

   Example: 4 was recorded on the first turn and then 5 was rolled. The player can't make 10 until a 5 or a 6 is rolled in future turns.

   $4 + \square = 10 \qquad 5 + 5 = 10 \qquad \square + \square = 10$

   $\square + \square = 10$

4. If "Choose" is rolled, the player may record any number (3–7) as an addend.
5. Players continue alternating turns and recording numbers whenever possible. When a number is rolled that cannot be recorded, the number cube is passed to the other player.
6. The first player to complete four equations wins. Both players may win by completing four equations in the same number of rolls.
7. To have children practice combining three addends, use the bottom half of the recording sheet with Number Cube (1–6).

Making Connections

Promote reflection and make mathematical connections by asking:

- Did you require more than 10 rolls to win this game? Please explain.
- Besides addition, what other skills are being practiced?

Roll Ten A

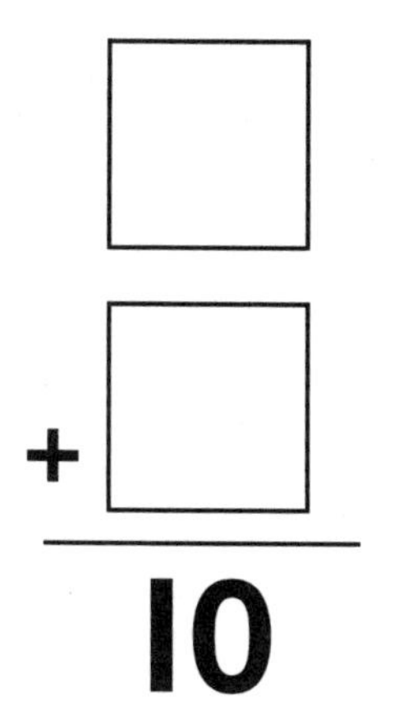

□	□	□
+ □	+ □	+ □
10	10	10

□ + □ = 10

Roll Ten B

□ + □ + □ = 10

□ + □ + □ = 10

□ + □ + □ = 10

□
□
+ □
10

Roll _____

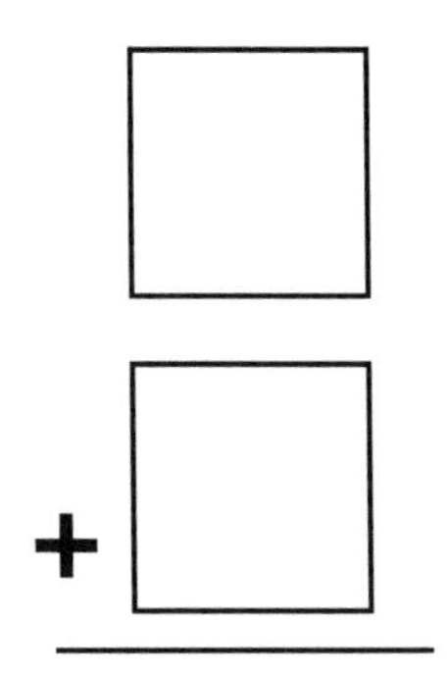

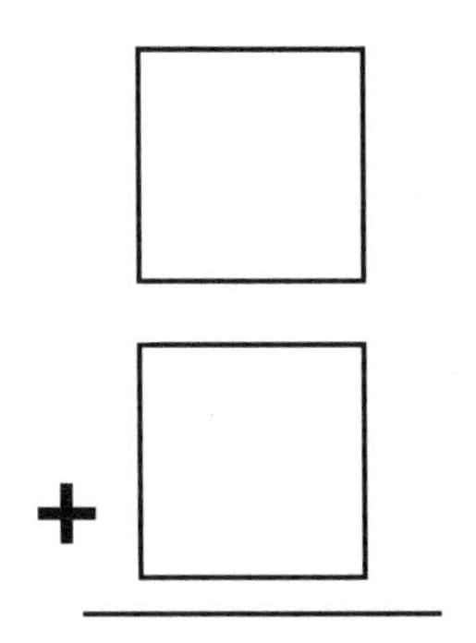

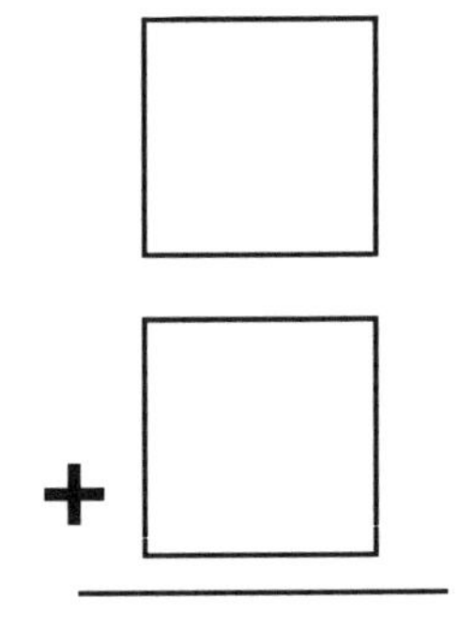

☐ + ☐ =

Roll _____

☐ + ☐ + ☐ =

☐ + ☐ + ☐ =

☐ + ☐ + ☐ =

Finding Addends

Topic: Addition Facts to 9

Object: Cover four numbers in a row with your counters.

Groups: Pair players or 2 players

Tip *Some children will benefit from the use of Dot Cards.*

Materials for each group

- *Finding Addends* gameboard, p. 50
- set of Digit Cards 4–9 (0–3 removed), p. 142
- counters (different kind for each pair)

Directions

1. The first pair mixes the Digit Cards, stacks them facedown, and draws one card. The pair finds and covers two adjacent numbers that total the number drawn. The drawn card is stacked in a discard pile.

   Example: If 6 is drawn, the pair might cover 5 and 1, 4 and 2, or 3 and 3.

8	4	1	4	2	4
1	3	5	3	6	3
5	2	6	2	3	6
6	3	4	4	5	1
1	3	5	3	4	4
7	6	2	2	3	2

2. The next pair draws a card, finds two adjacent numbers that total the number, and covers the two adjacent numbers with distinct counters. The pair discards the drawn card.
3. Pairs continue alternating turns and trying to arrange four of their counters in a row.
4. After six Digit Cards are drawn, one player mixes the Digit Cards and stacks them facedown for continued use.
5. If a pair draws a card and cannot find two addends, the pair passes that turn.
6. The game ends when one pair lines up four of that pair's counters in a row, or when all adjacent addends are covered without a four-in-a-row arrangement.

Making Connections

Promote reflection and make mathematical connections by asking:

- What strategy helped you line up your counters in a row?
- How would you play this game differently next time?

Finding Addends

8	4	1	4	2	4
1	3	5	3	6	3
5	2	6	2	3	6
6	3	4	4	5	1
1	3	5	3	4	4
7	6	2	2	3	2

Choose Two

Topic: Addition Facts to 12

Object: Cover three numbers in a row with your counters.

Groups: Pair players or 2 players

Materials for each group

- *Choose Two* gameboard, p. 52
- 3 Number Cubes (1–6), p. 147
- counters (different kind for each pair)

Tips *Some children will benefit from the use of Dot Cubes. Extend the playing time and promote more strategic thinking by requiring four-in-a-row.*

Directions

1. The first pair rolls the three Number Cubes, chooses two of the three rolled numbers to add, states the addition equation, and covers the corresponding sum and addends on the gameboard.
2. If two of the rolled numbers cannot add to five or more, the pair rolls again.
3. The other pair rolls three Number Cubes and follows the same procedure.

   Example: If 4, 3, and 5 are rolled, the pair might cover 7 (4 + 3), 8 (3 + 5), or 9 (4 + 5).
4. When none of the rolled numbers match the uncovered addends on the gameboard, the pair may cover a matching sum with different addends.

   Example: If 4, 3, and 5 are rolled and all the above choices are covered, pair could cover 7 (5 + 2) or (6 + 1).
5. If all possible sums are already covered, the pair loses that turn.
6. The first pair to have three counters in a row horizontally, vertically, or diagonally wins.

8 3 + 5	7 4 + 3	5 2 + 3	12 6 + 6	6 5 + 1
7 6 + 1	10 5 + 5	9 6 + 3	6 1 + 5	7 2 + 5
6 3 + 3	8 6 + 2	7 1 + 6	8 4 + 4	9 5 + 4
11 6 + 5	7 3 + 4	8 5 + 3	10 4 + 6	6 4 + 2
9 4 + 5	5 4 + 1	6 2 + 4	8 2 + 6	7 5 + 2

Making Connections

Promote reflection and make mathematical connections by asking:

- How many different sums can you make from three different rolled numbers? What if two of the numbers are the same?
- What strategies helped you line up your markers in a row?

Choose Two

8 3 + 5	**7** 4 + 3	**5** 2 + 3	**12** 6 + 6	**6** 5 + 1
7 6 + 1	**10** 5 + 5	**9** 6 + 3	**6** 1 + 5	**7** 2 + 5
6 3 + 3	**8** 6 + 2	**7** 1 + 6	**8** 4 + 4	**9** 5 + 4
11 6 + 5	**7** 3 + 4	**8** 5 + 3	**10** 4 + 6	**6** 4 + 2
9 4 + 5	**5** 4 + 1	**6** 2 + 4	**8** 2 + 6	**7** 5 + 2

Date ____________________ Name ________________________________

Seeking Sums Practice I

Color two amounts to equal the sum in the triangle.

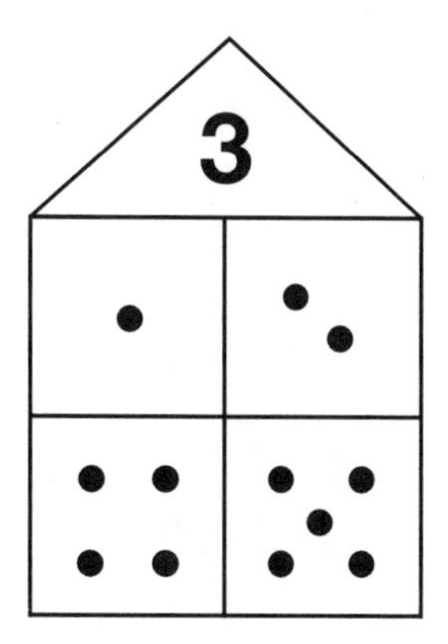

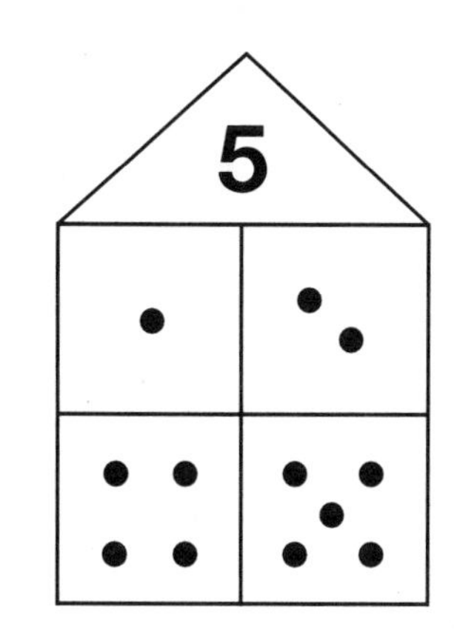

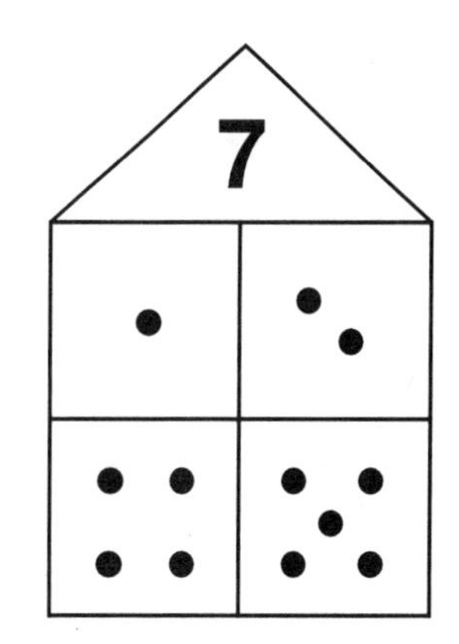

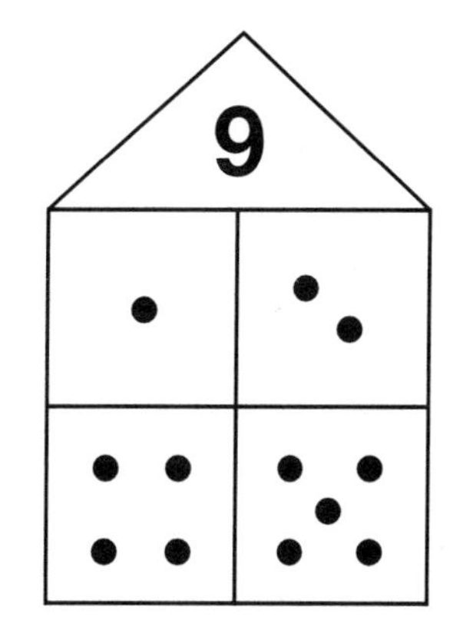

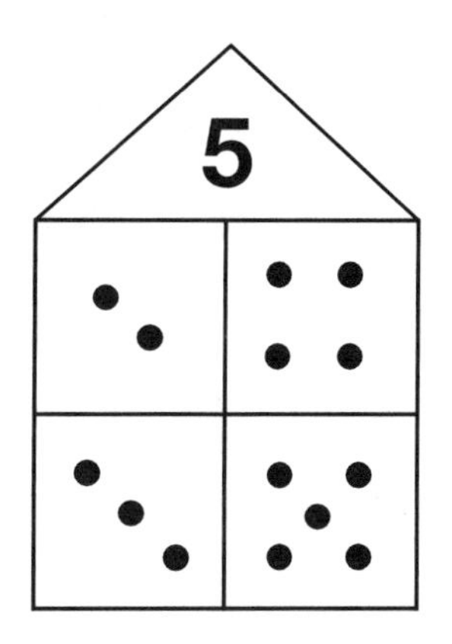

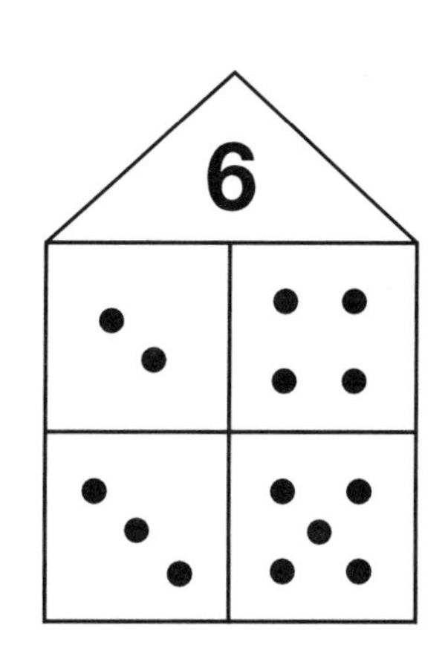

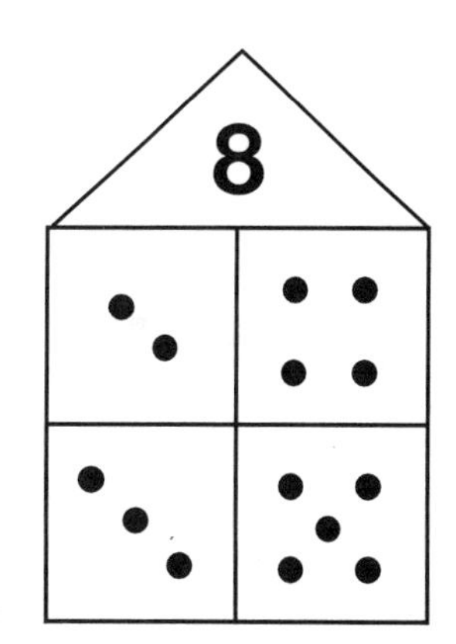

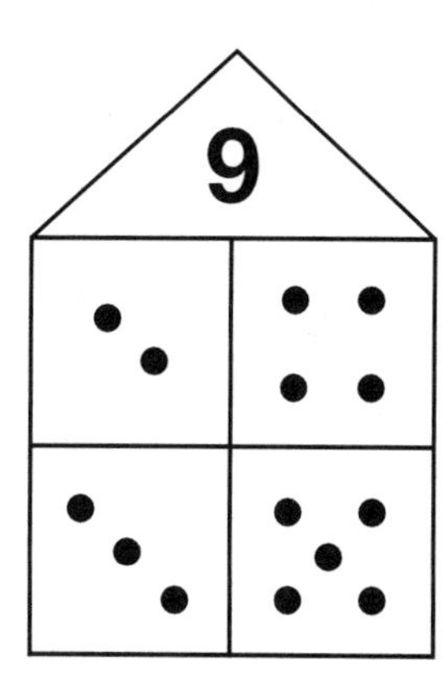

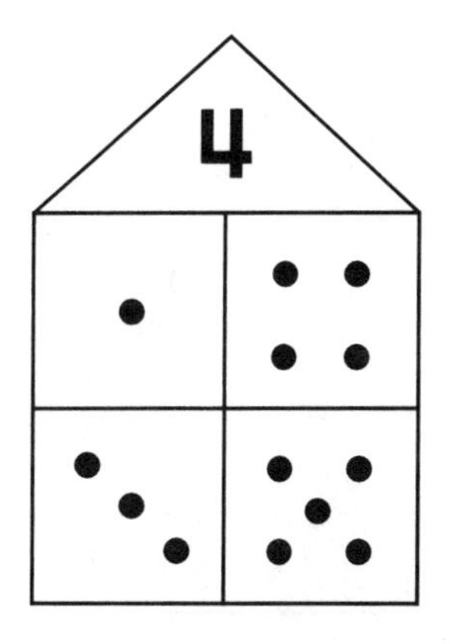

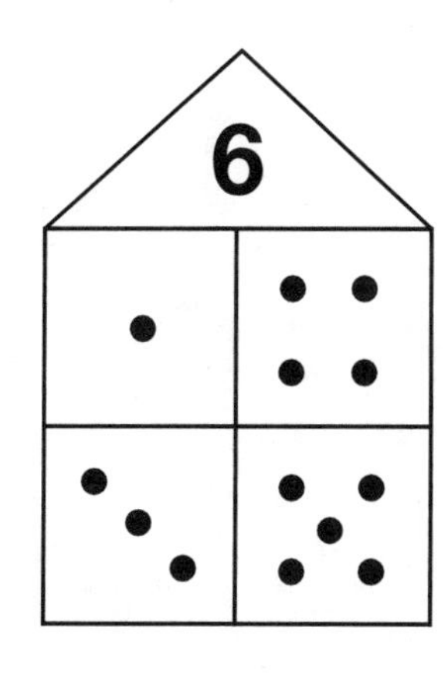

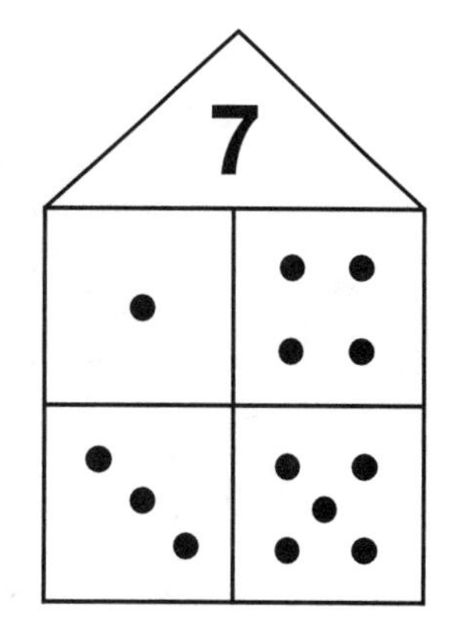

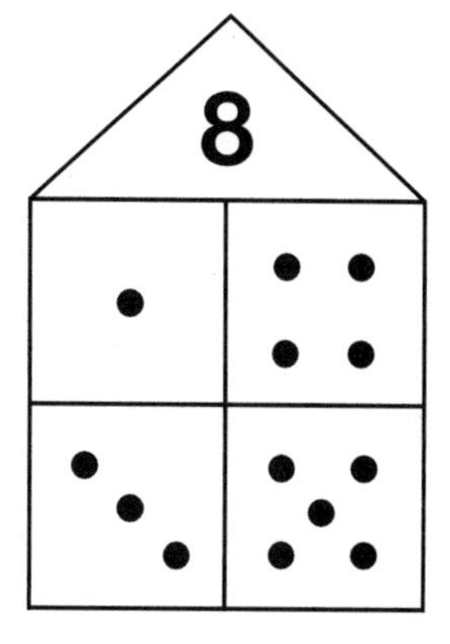

Date ______________ Name ______________________

Seeking Sums Practice II

Color two or three amounts to equal the sum in the triangle.

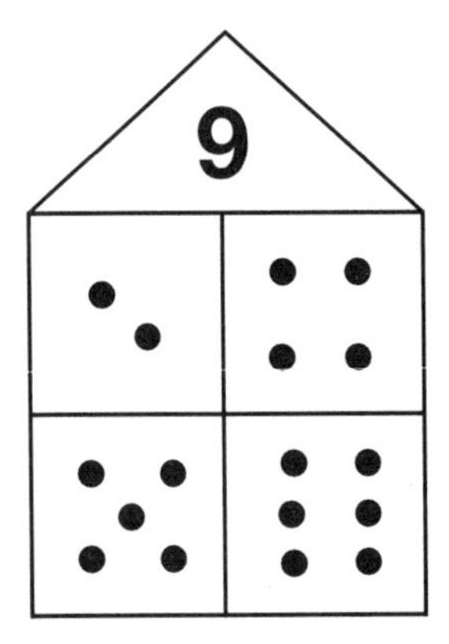

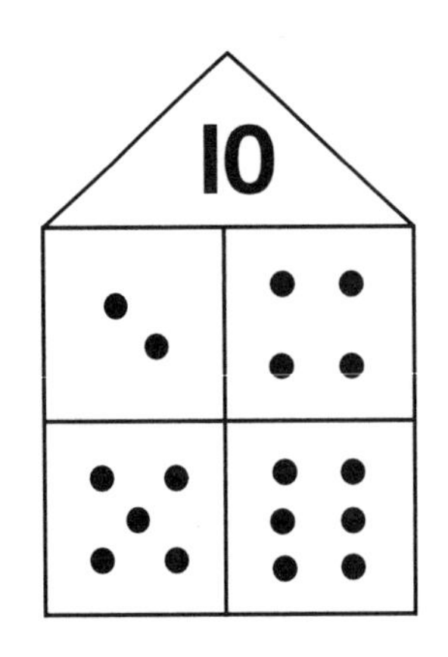

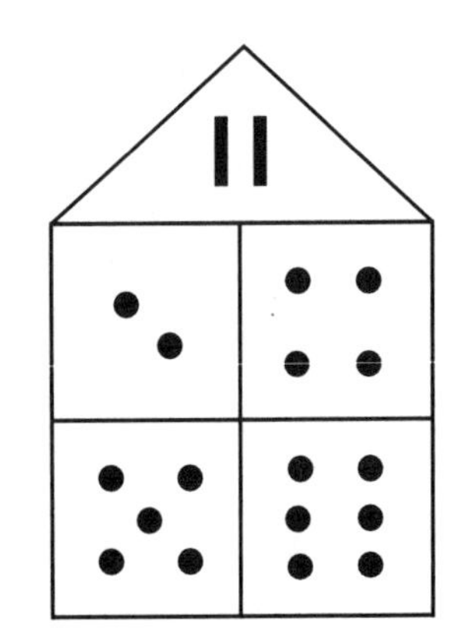

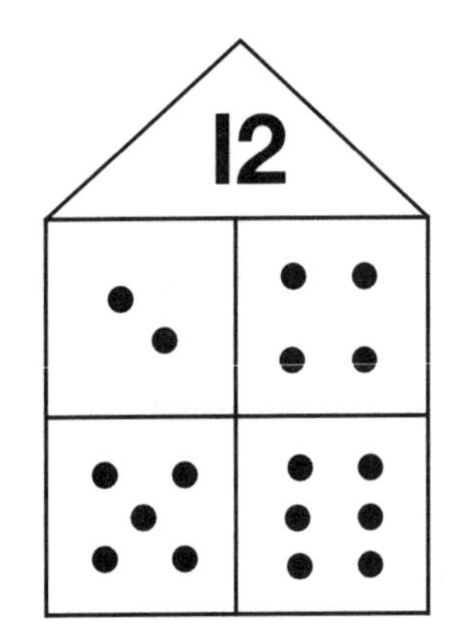

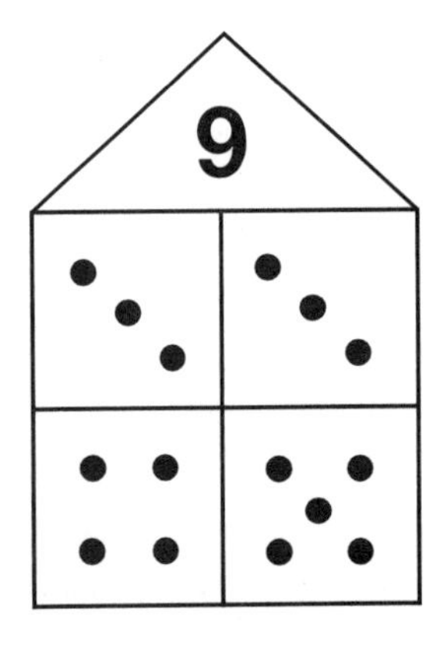

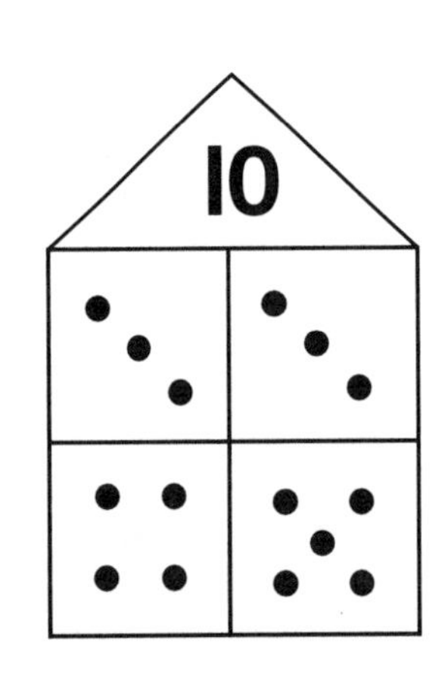

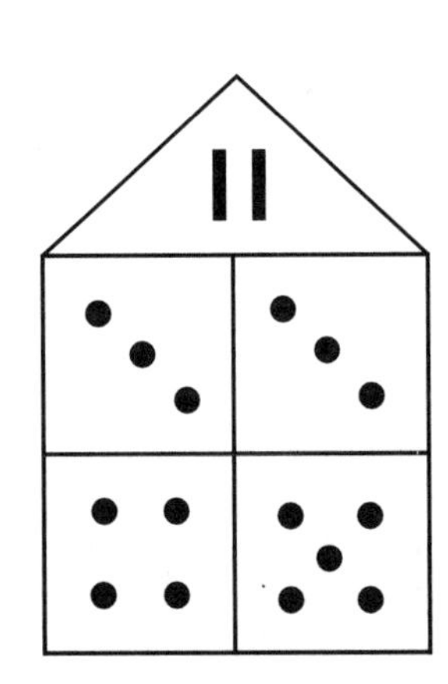

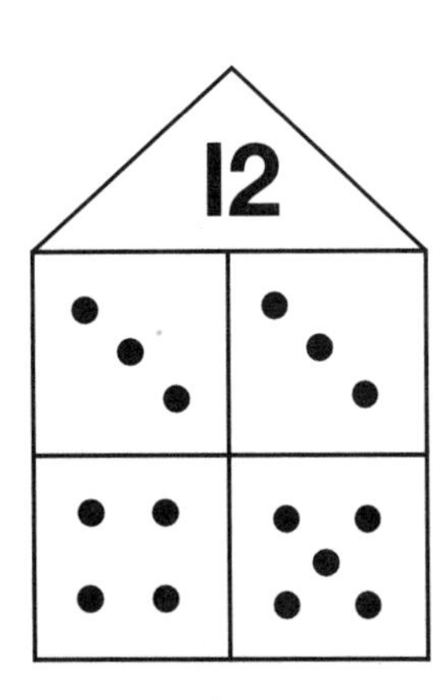

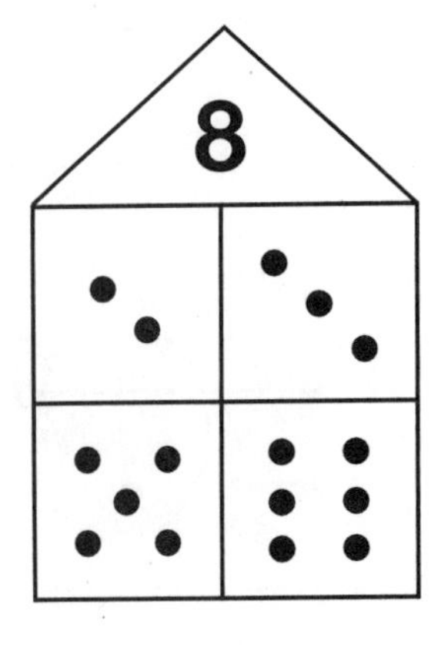

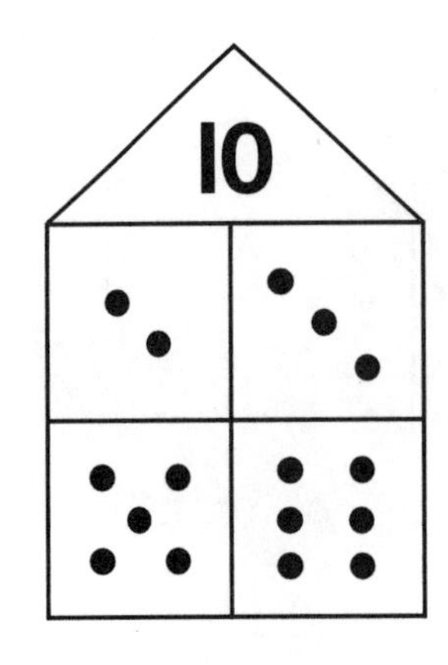

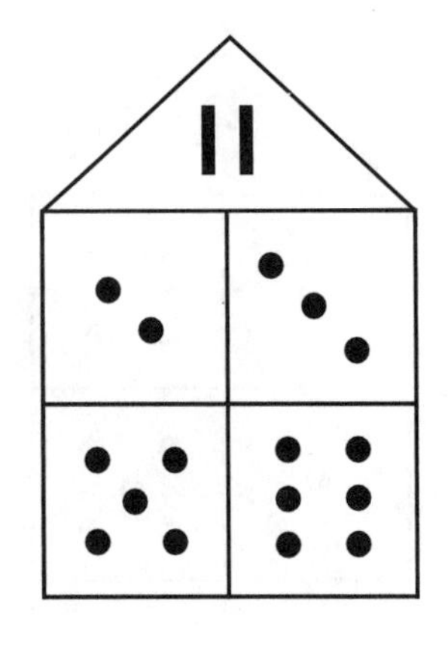

13

Date ____________________ Name ________________________________

Roll and Fill

Roll a number cube to fill in at least one number for each problem. Will you roll some of the second numbers?

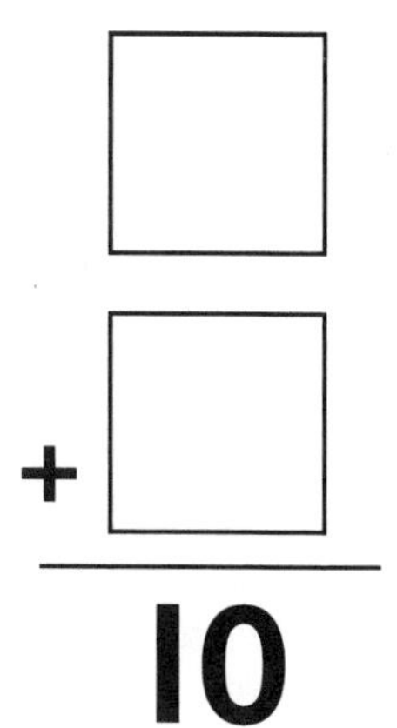
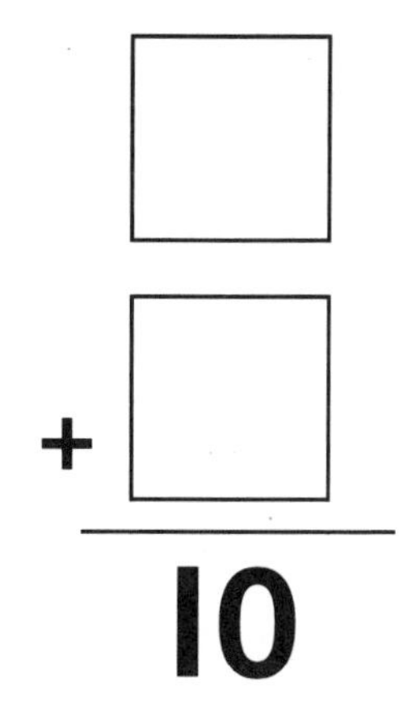
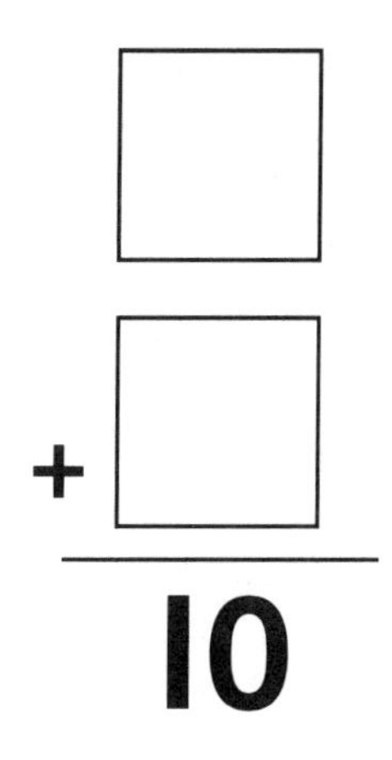

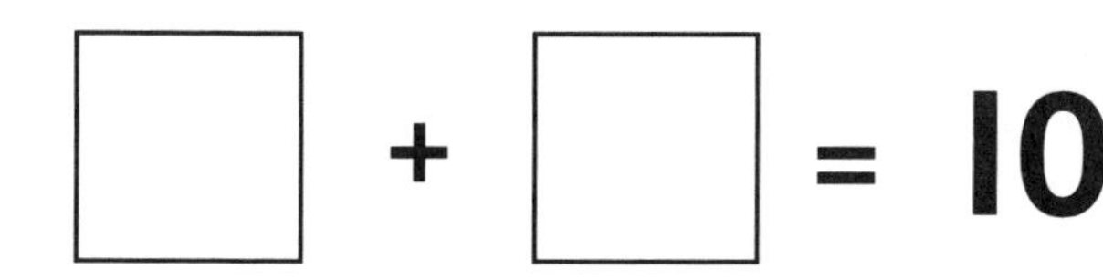

Roll a number cube to fill in at least two of the numbers for each problem. How many third numbers can you roll?

☐ + ☐ + ☐ = 12

☐ + ☐ + ☐ = 12

☐ + ☐ + ☐ = 12

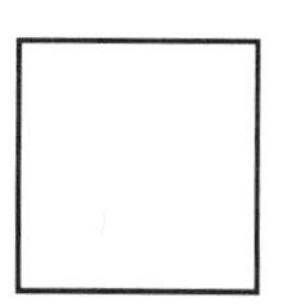

☐
+ ☐

12

If time runs out, fill in any numbers to correctly complete the equations.

Date ____________ Name ____________________

Roll and Fill

Roll a number cube to fill in at least one number for each problem. Will you roll some of the second numbers?

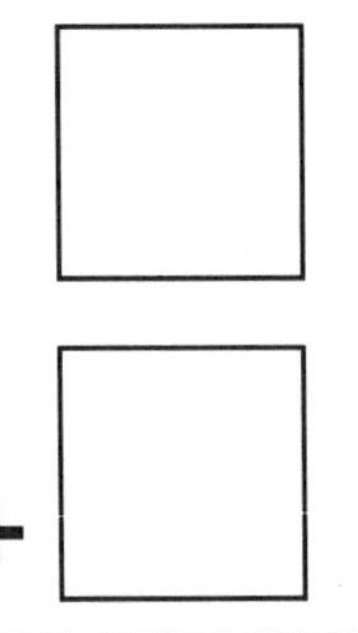

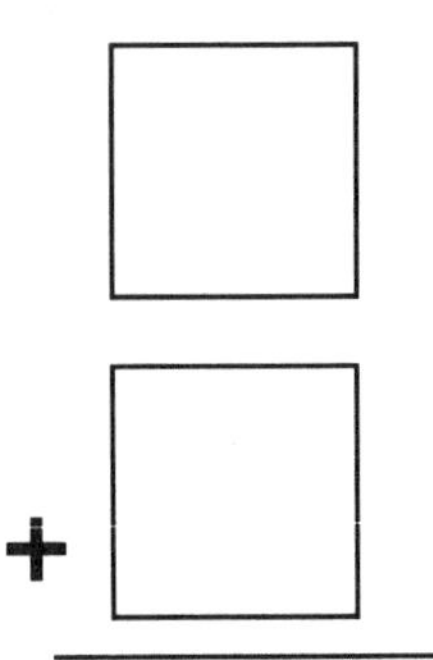

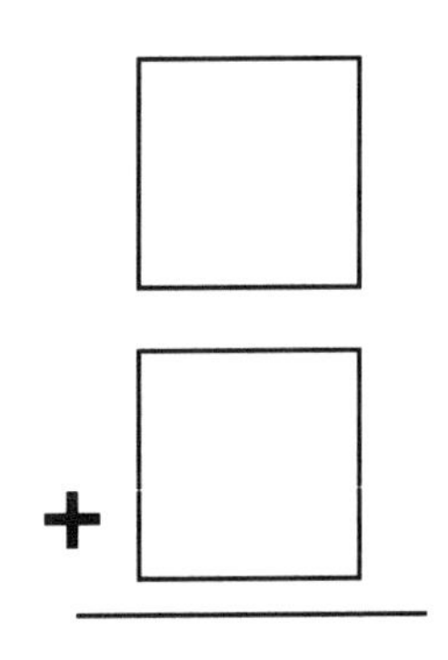

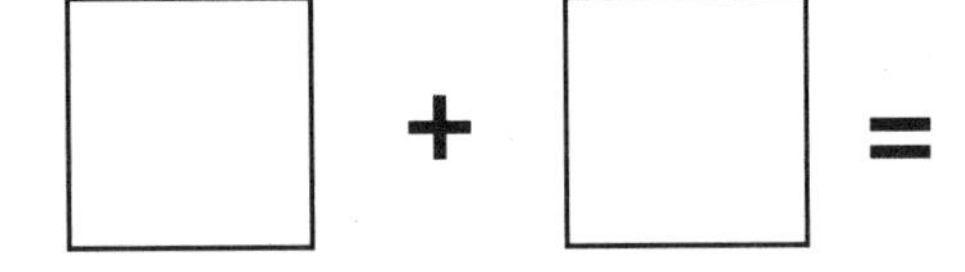

Roll a number cube to fill in at least two of the numbers for each problem. How many third numbers can you roll?

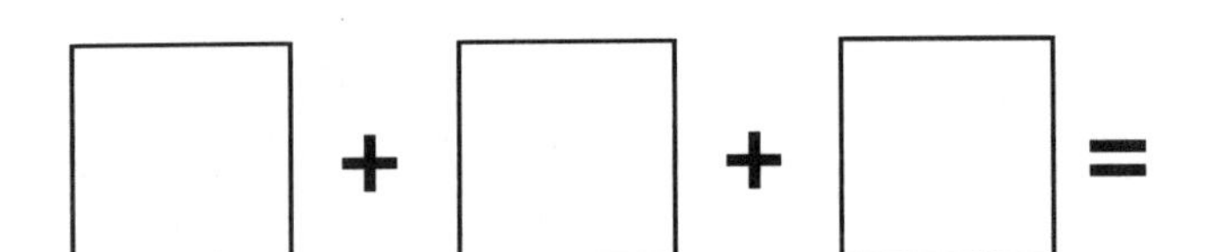

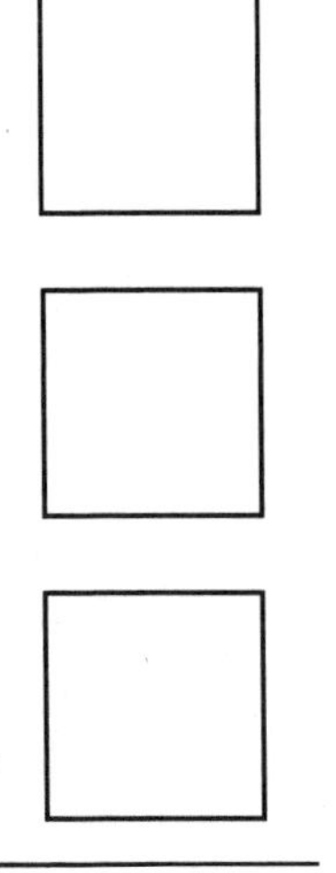

If time runs out, fill in any numbers to correctly complete the equations.

Note to teacher: Identify and record sums before duplicating.

Date ____________ Name ____________

Neighbor Sums I

Circle the numbers next to each other to equal the sum shown.

Sum 6

3	3	4	5	1	4	2

3	4	4	2
3	5	1	6

Sum 7

4	4	3	5	2	6	1

2	5	3	3
4	1	6	4

Sum 8

2	4	4	1	3	5	2

6	2	4	5	1	7	3

5	2	6	5
6	3	4	3
1	7	4	6

Sum 9

5	3	5	4	2	8	1

4	6	3	5	7	2	6

6	1	7	2
2	8	4	6
6	3	5	2

Date ______________ Name ______________________

Neighbor Sums II

Circle the numbers next to each other to equal the sum shown.

Sum 9

5	3	6	4	4	5	3

3	8	2	7	3	1	8

5	4	3	2
6	4	6	7
3	6	8	1

Sum 10

3	8	2	5	4	6	3

9	1	6	5	5	3	7

4	6	3	2
1	5	5	8
6	4	7	3

Sum 11

2	8	3	7	3	9	2

7	4	9	2	6	5	7

9	2	8	2
3	7	3	7
8	5	6	4

Sum 12

2	8	4	7	6	6	5

6	4	7	5	8	3	9

9	3	8	3
4	5	4	7
8	6	6	5

Independent Activity

Subtraction Facts

Assumptions The subtraction facts have previously been taught and reviewed with an emphasis on understanding. Concrete objects and visual models, such as counters and Ten Frames, have been used extensively.

Section Overview and Suggestions

Sponges

Ten Frame Differences p. 60

Disappearing Robot p. 61

Seeking Differences pp. 62–63

These open-ended, whole-class, or small-group warm-ups actively engage children in practicing many subtraction facts. The frequent use of these Sponges will ensure greater success with the Games and Independent Activities in this section.

Skill Checks

Just the Facts 7–12 pp. 64–66

These provide a way to help parents, children, and you see children's improvement with the subtraction facts. Copies of *Just the Facts* may be cut in half so that each Check may be used at a different time. Remember to have children respond to STOP, the number sense task, before they solve the ten problems.

Games

Taking from Ten pp. 67–68

Choose and Subtract pp. 69–71

How Many More? pp. 72–74

These open-ended, repeatable Games actively involve children in practicing many subtraction facts. Using a Ten Frame to visualize amounts, pairs work cooperatively rather than competitively to complete *Taking from Ten. Choose and Subtract,* a challenging Game, encourages flexible thinking. In *How Many More?* children practice a worthwhile strategy of "counting on" to subtract. All Games promote mental computation as children enhance their strategic thinking skills.

Independent Activities

Seeking Differences Practice pp. 75–76

Roll and Fill Differences pp. 77–78

Equation Hunt pp. 79–80

Each Independent Activity has two worksheets that allow children to independently practice the easier and the more difficult subtraction facts. *Seeking Differences Practice* uses the helpful visual dot patterns format. Some problems have more than one solution. *Roll and Fill Differences* includes open-ended, repeatable recording sheets that allow children to practice varied facts. *Equation Hunt* requires children to seek three numbers to create a subtraction equation.

Ten Frame Differences

Tip When appropriate, use two Ten Frames to practice more challenging subtraction facts. Example: Make 12. Visualize 4 less.

Topic: Subtraction Facts

Object: Visualize, verbalize, and display differences.

Groups: Whole class or small group

Materials

- Ten Frame form for each child, p. 149
- transparency of Ten Frame form, p.149
- counters (for each child)
- Number Cube (1–6), p. 147

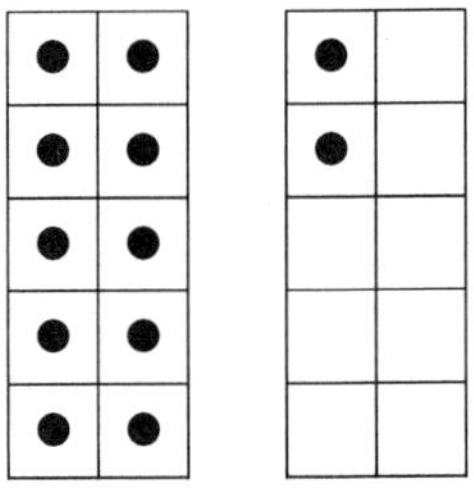

Directions

1. The leader places seven or more counters in the displayed Ten Frame and asks children, "How many?"
2. Children display the same number of counters in their individual Ten Frames.
3. One child rolls the Number Cube and announces the amount.
4. Using the rolled amount, the leader asks children to visualize removing that many counters from the Ten Frame, and identify the new difference.

   Example: If 8 is displayed on a Ten Frame and a 3 is rolled, children visualize 3 counters being removed from the Ten Frame.

5. After children state the difference and equation, children display the difference on their Ten Frame, placing the removed counters next to the Ten Frame. The leader should draw attention to how the two displayed parts equal the original amount. 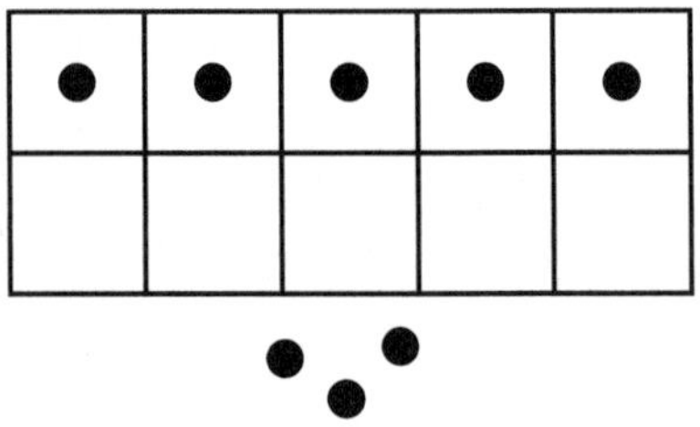
6. Children clear their Ten Frames and repeat steps 1–5.

Making Connections

Promote reflection and make mathematical connections by asking:

- How did the Ten Frame help you identify the correct difference?
- Which amounts are easier to visualize? Why?

Disappearing Robot

Topic: Subtraction Facts

Object: Erase robot parts by solving subtraction equations.

Groups: Whole class or small group

Tip *Numbers written inside the robot parts can be varied according to the range of child abilities. With younger children, you could fill the parts with 2, 3, and 4 only.*

Materials

- chalkboard and chalk
- chalk eraser

Directions

1. The leader draws a robot with numbered parts (similar to illustration) on the chalkboard. Numbers one through six work best.
2. The leader states a subtraction fact that equals a number found inside a robot part.
3. The leader selects a child to state the completed subtraction equation. If correct, the child erases a robot part containing the difference.

   Example: "What's the difference between 10 and 5?" Responder: "10 minus 5 equals 5." (Responder erases the hand.)
4. The leader continues to state subtraction facts and select children to respond.

   Example: "What's left when you subtract 4 from 7?" Responder: "7 minus 4 equals 3." (Responder erases the ear containing 3.)
5. Use subtraction situations in context to vary this activity.

   Example: "Cindy has 6 dimes and needs 10 dimes to buy a gift. How many more dimes does she need?" Responder: "6 and 4 make 10 because 10 minus 6 equals 4." (Responder erases a part containing a 4.)
6. If appropriate, responding children can provide subtraction facts for others to answer.

Making Connections

Promote reflection and make mathematical connections by asking:

- How did you get that answer? Please explain.

Seeking Differences

Topic: Subtraction Facts

Object: Select dot cards to equal target differences.

Groups: Whole class or small group

Tip *Transition into Digit Cards as children seem ready.*

Materials

- transparent Dot Pattern Cards, p. 145
- transparency of *Seeking Differences* activity form, p. 63
- 8 transparent counters

Directions

1. The leader randomly selects four Dot Pattern Cards and displays them on the *Seeking Differences* activity form.
2. Using any of the four displayed amounts, children identify ways to make the differences one through eight.

   Example: If 1, 2, 4, and 7 are displayed, 3 could be made two different ways (7 – 4 or 4 – 1).
3. When children identify a way to show a difference, the leader covers the difference found at the bottom of the activity form. The leader should encourage players to seek additional solutions for the same difference.
4. Even though some differences are not possible, encourage children to seek possible solutions.

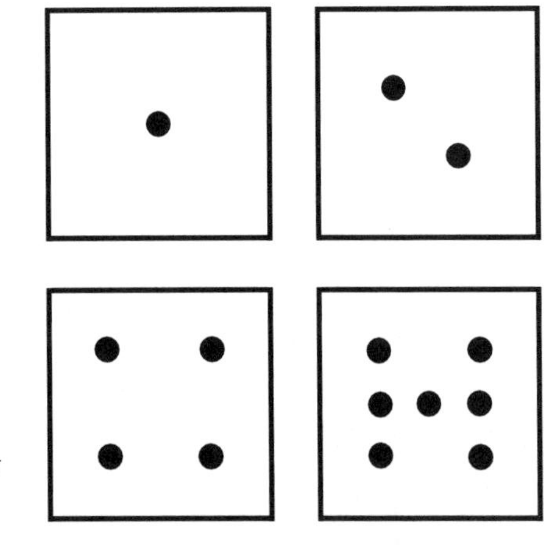

1	2	3	4
5	6	7	8

Making Connections

Promote reflection and make mathematical connections by asking:

- Which differences were not possible? If so, why not?
- Which differences could be made more than one way?
- What dot patterns might allow for more differences?

Seeking Differences

1	2	3	4
5	6	7	8

Date ____________ Name ____________________

Just the Facts 7

Don't start yet! Circle the problem in the top row that may have the smallest answer.

1. 5 – 2 = ______ **2.** 7 – 5 = ______ **3.** 8 – 2 = ______

4. 7 – 1 = ____ **5.** 8 – 6 = ____ **6.** 11 – 5 = ____ **7.** 13 – 6 = ____

8. 10 – 6 = ______ **9.** 2 + ______ = 4 **10.** 5 + ______ = 6

Go On What number is missing? 11, 9, 7, ________, 3, 1
What is the subtraction rule?

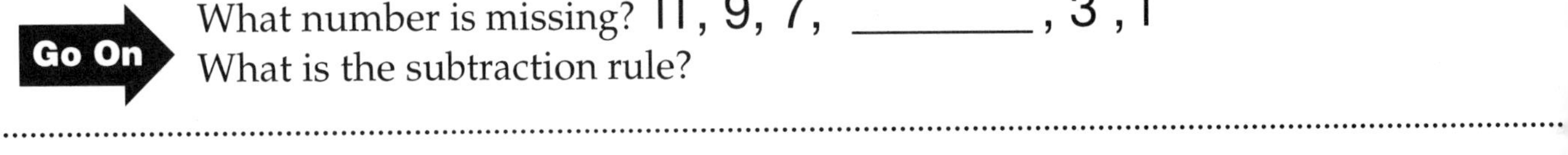

Date ____________ Name ____________________

Just the Facts 8

Don't start yet! Circle the problem in the top row that may have the largest answer.

1. 5 – 4 = ______ **2.** 7 – 4 = ______ **3.** 9 – 2 = ______

4. 6 – 1 = ____ **5.** 9 – 4 = ____ **6.** 12 – 6 = ____ **7.** 14 – 7 = ____

8. 10 – 3 = ______ **9.** 1 + ______ = 2 **10.** 6 + ______ = 7

Go On Write three subtraction facts that equal 1.

Date ________________ Name ________________________

Just the Facts 9

Don't start yet! Circle two problems that may have answers less than 3.

1. 5 – 3 = ______ **2.** 9 – 6 = ______ **3.** 6 – 2 = ______

4. 6 – 5 = ______ **5.** 8 – 5 = ______ **6.** 11 – 7 = ______ **7.** 13 – 9 = ______

8. 10 – 8 = ______ **9.** 3 + ______ = 6 **10.** 4 + ______ = 5

What numbers are missing? 12, 10, 8, ______, ______, 2
What is the subtraction rule?

Date ________________ Name ________________________

Just the Facts 10

Don't start yet! Circle two problems that may have answers greater than 5.

1. 6 – 5 = ______ **2.** 8 – 3 = ______ **3.** 10 – 2 = ______

4. 7 – 6 = ______ **5.** 9 – 2 = ______ **6.** 12 – 5 = ______ **7.** 14 – 9 = ______

8. 10 – 9 = ______ **9.** 4 + ______ = 8 **10.** 8 + ______ = 9

Go On Write three subtraction facts that equal 2.

Date ____________ Name ____________________

Just the Facts 11

Don't start yet! Circle the problem in the top row that may have the smallest answer.

1. 6 – 3 = ______ **2.** 8 – 4 = ______ **3.** 7 – 2 = ______

4. 6 − 6 ___ **5.** 8 − 7 ___ **6.** 11 − 8 ___ **7.** 13 − 5 ___

8. 10 – 7 = ______ **9.** 5 + ______ = 10 **10.** 9 + ______ = 10

What number is missing? 13, 10, 7, ______, 1

What is the subtraction rule?

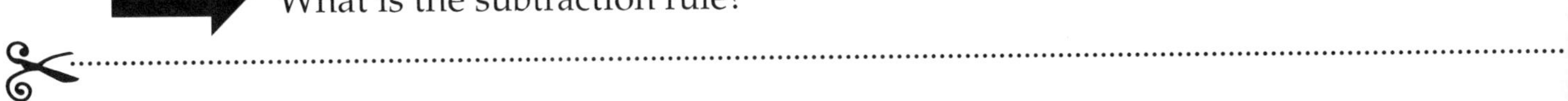

Date ____________ Name ____________________

Just the Facts 12

Don't start yet! Circle the problem in the top row that may an answer greater than 5.

1. 6 – 4 = ______ **2.** 9 – 7 = ______ **3.** 11 – 2 = ______

4. 7 − 3 ___ **5.** 9 − 5 ___ **6.** 12 − 8 ___ **7.** 14 − 6 ___

8. 10 – 5 = ______ **9.** 6 + ______ = 12 **10.** 7 + ______ = 8

Go On Write three subtraction facts that equal 3.

Taking from Ten

Topic: Subtraction Facts

Object: Fill a pathway to the star.

Groups: Pairs

Materials for each group

- *Taking from Ten* gameboard, p. 68
- Ten Frame, p. 149
- Dot Cube (1-2-3-4-5-Choose), p. 144
- 35 counters

Tips *As pairs play, one child might record the corresponding equations. Place the removed counters close to the Ten Frame so that children can still see the two parts.*

Directions

1. In this game, a pair works cooperatively to fill a pathway.
2. Fill the Ten Frame with 10 counters.
3. The pair rolls the Dot Cube. If "Choose" is rolled, the pair may select any number one through five. When a number is rolled, the pair removes that amount of counters from the Ten Frame, states the equation, and places a counter on the path with that difference.

   Example: If 4 is rolled, 4 counters are removed from the Ten Frame, "10 minus 4 equals 6" is stated, and a counter is placed in the first cell of the 6-pathway.
4. The pair continues until a path to the star is filled. (To extend the playing time, have children attempt to complete two or more pathways.)

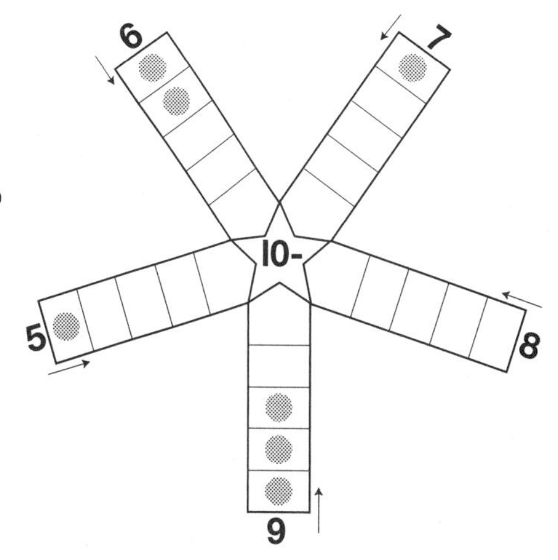

Making Connections

Promote reflection and make mathematical connections by asking:

- How did the Ten Frame help you identify the correct difference and pathway?

Taking from Ten

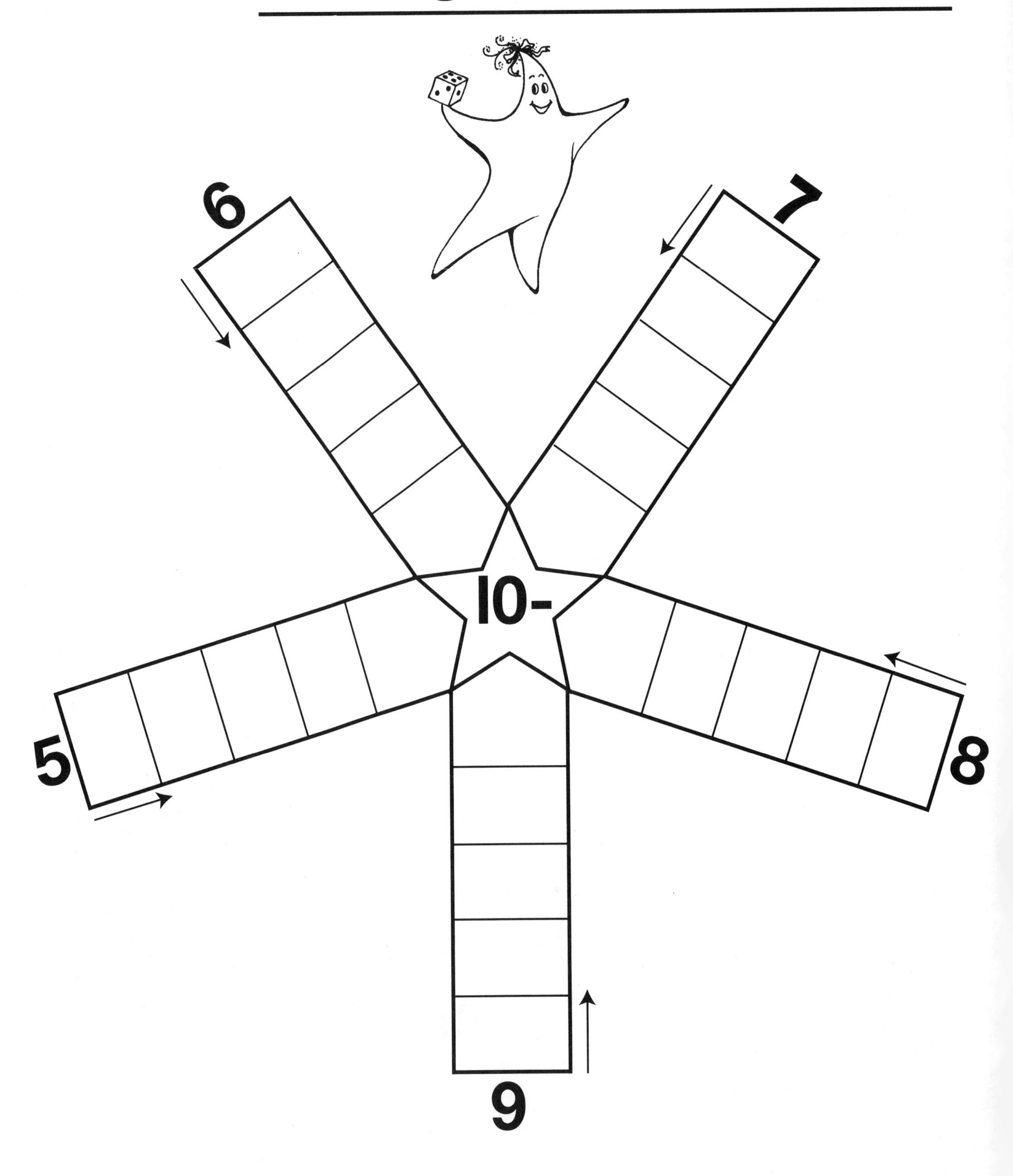

Choose and Subtract

Topic: Subtraction Facts

Object: Cover three numbers in a row with your counters.

Groups: Pair players or 2 players

Materials for each group

- *Choose and Subtract A* gameboard, p. 70
- 3 Number Cubes (1–6), p. 147
- counters (different kind for each pair)

***Tips** Some children will benefit from use of Dot Cubes. Extend the playing time and promote more strategic thinking by requiring four-in-a-row.*

Directions

1. The first pair rolls the three Number Cubes.
2. The pair selects two of the three Number Cubes to subtract, states the subtraction equation, and places a counter on the resulting difference.

   Example: If 2, 3, and 5 are rolled, pair could cover 1 (3 – 2), 2 (5 – 3), or 3 (5 – 2).
3. Pairs alternate turns rolling Number Cubes, stating equations, and placing counters on the gameboard.
4. If all the resulting differences are already covered, the pair loses that turn.
5. The first pair to have three counters in a row horizontally, vertically, or diagonally wins.
6. To have children practice more difficult subtraction facts, use *Choose and Subtract B* gameboard with two special Number Cubes (5–10) and one Number Cube (1–6).

2	1	3	2	0
0	3	2	4	1
1	0	3	1	3
5	2	0	2	1
1	4	2	1	0

Making Connections

Promote reflection and make mathematical connections by asking:

- How many differences can you make from three different rolled numbers? Give examples.
- What strategies helped you line up your counters in a row?

Choose and Subtract A

2	1	3	2	0
0	3	2	4	1
1	0	3	1	3
5	2	0	2	1
1	4	2	1	0

Choose and Subtract B

0	3	4	6	1
2	5	1	2	9
1	3	4	1	0
7	0	6	0	2
3	8	2	5	4

How Many More?

Topic: Subtraction Facts

Object: Cover three in a row with your markers.

Groups: Pair players

Tip *Substitute Dot Cubes or Dot Pattern Cards (1–6) if children are not ready for the number cube.*

Materials for each group

- *How Many More to Make 8?* gameboard, p. 73
- counters (different kind for each pair)
- Number Cube (1–6), p. 147
- paper for recording equations

Directions

1. The first pair rolls the Number Cube to determine how many more are needed to make 8. The pair places a counter on a dot pattern that represents the missing amount.

   Example: If 2 is rolled, 6 is needed to make 8. Thus, the pair selects and covers one of the six dot patterns on the gameboard.
2. Pairs are required to say aloud the related subtraction fact for each turn. (If 3 is rolled, the pair states "I have 3. I need 5 more to make 8 because 8 minus 3 equals 5.")
3. Pairs alternate turns following this procedure. The first pair to place three of their counters in a row horizontally, vertically, or diagonally wins.
4. To provide practice with 10, use the *How Many More to Make 10?* gameboard, p. 74.

Making Connections

Promote reflection and make mathematical connections by asking:

- Were some numbers easier to cover than others?
- What strategies did you use in placing your counters?
- Was it difficult to block your opponent? Why or why not?

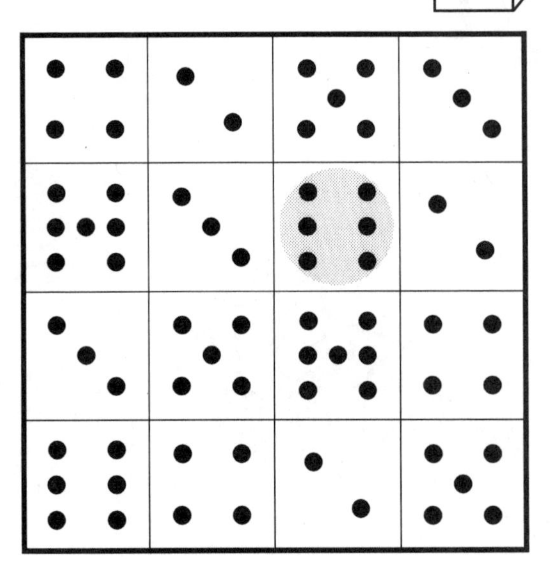

How Many More to Make 8?

How Many More to Make 10?

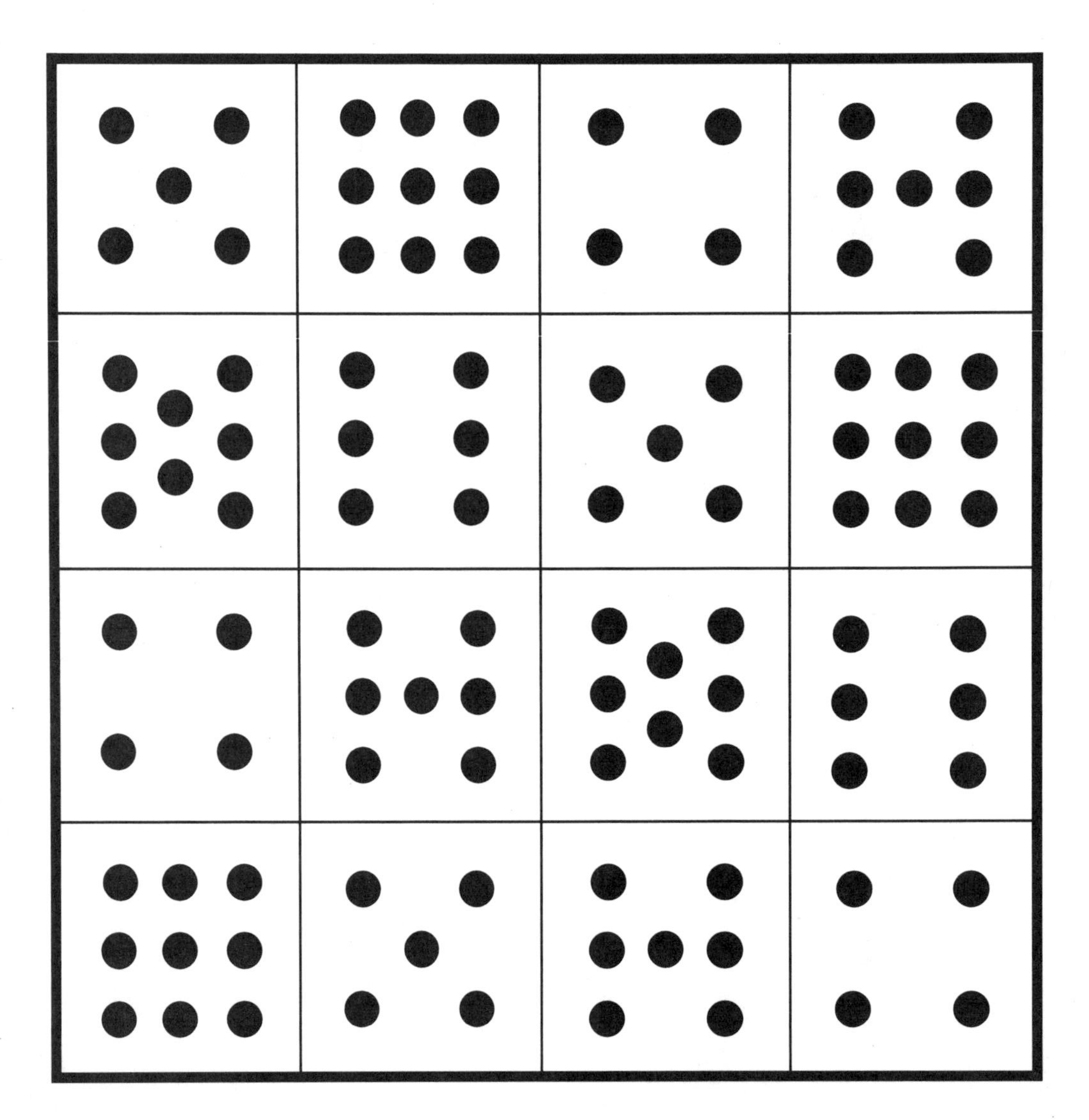

Date ______________ Name ______________________

Seeking Differences Practice I

Color two amounts to equal the difference in the triangle.

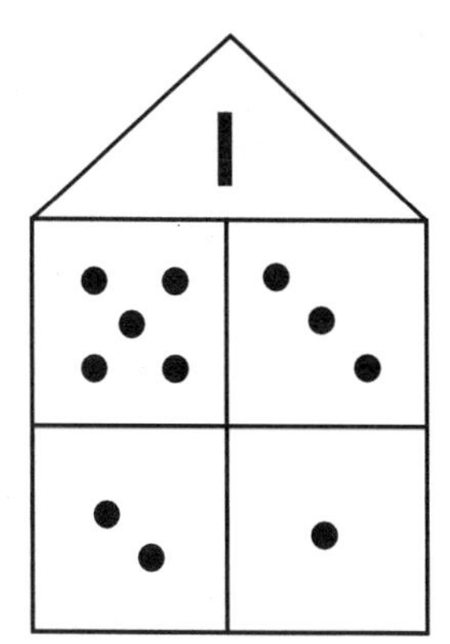

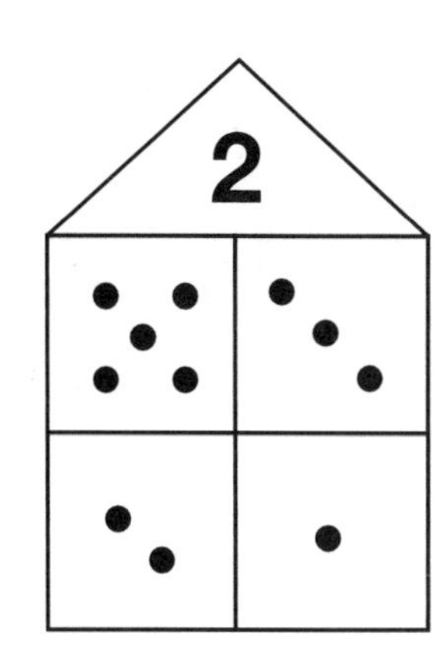

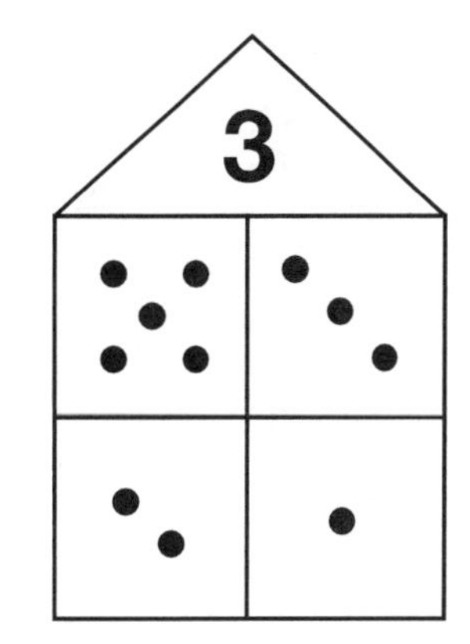

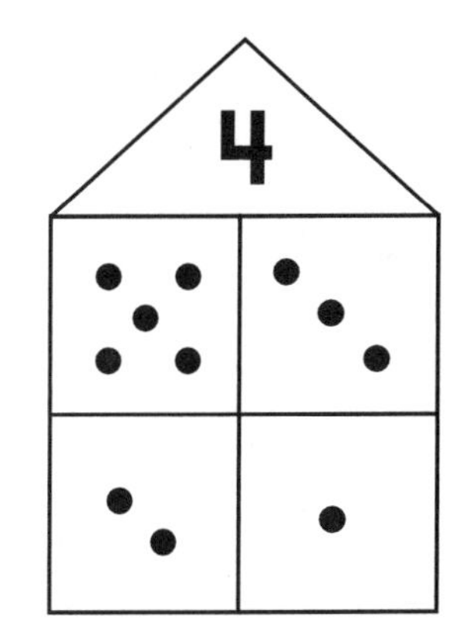

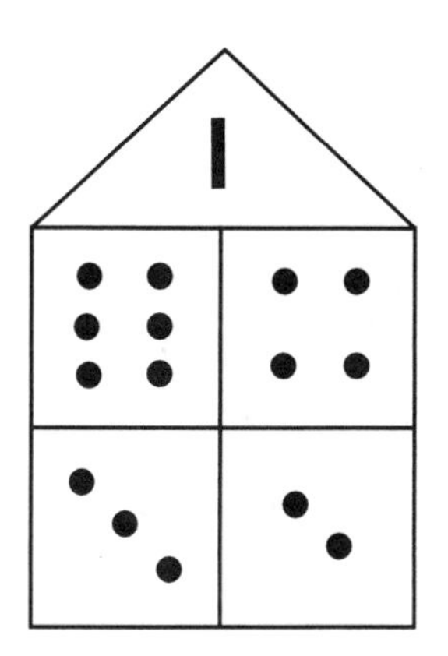

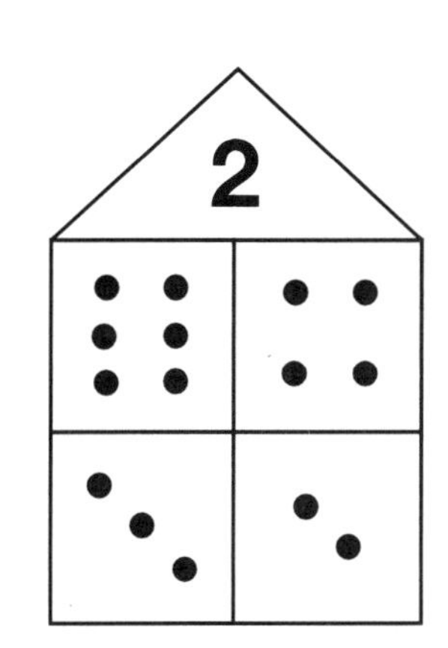

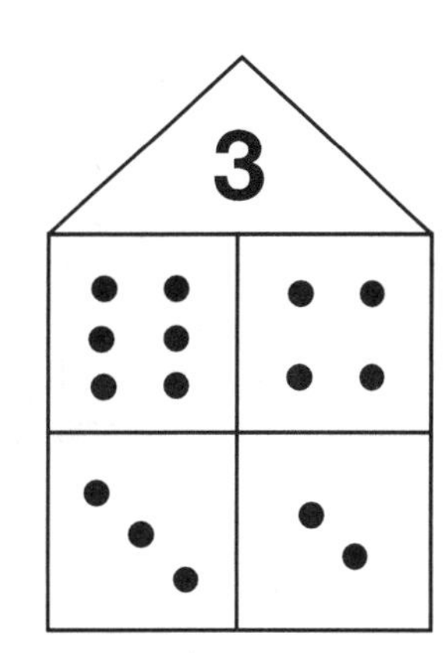

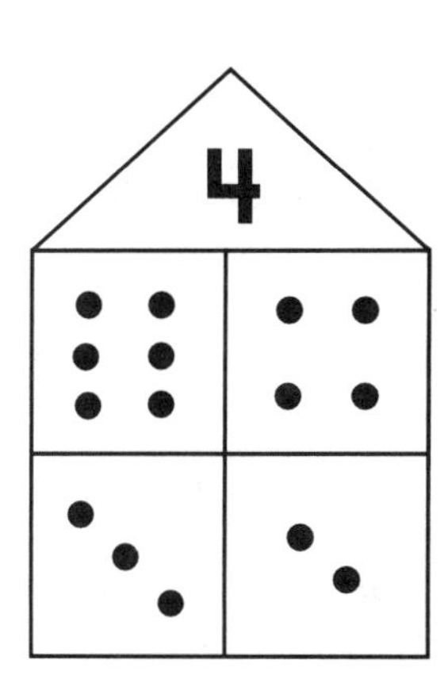

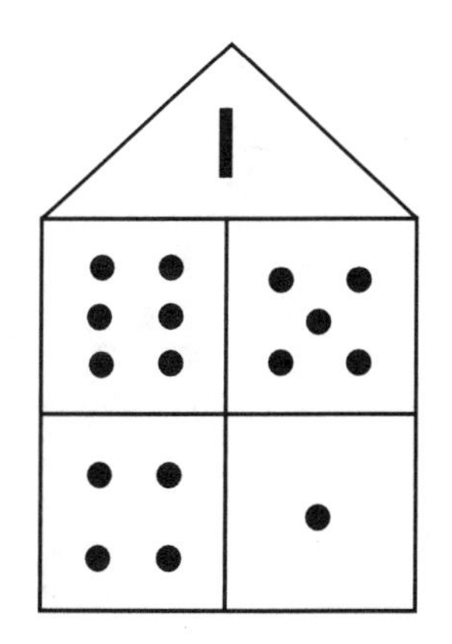

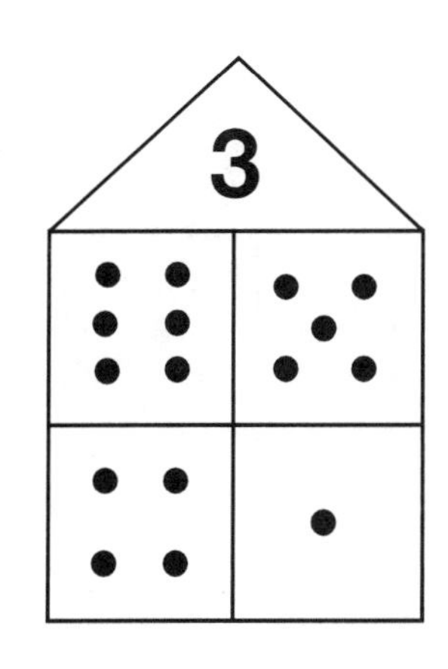

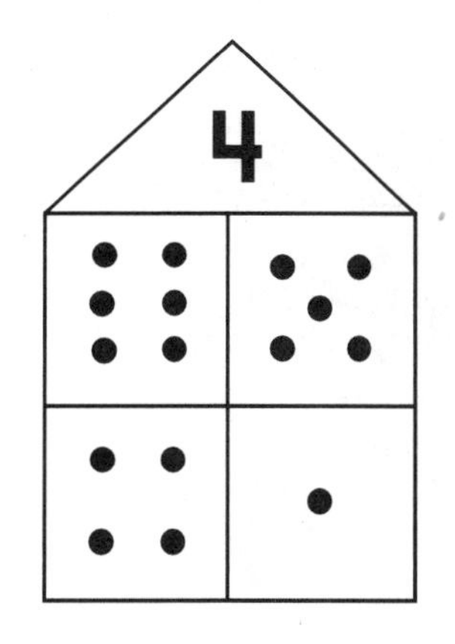

5

Date ____________________ Name ____________________________________

Seeking Differences Practice II

Color two amounts to equal the difference in the triangle.

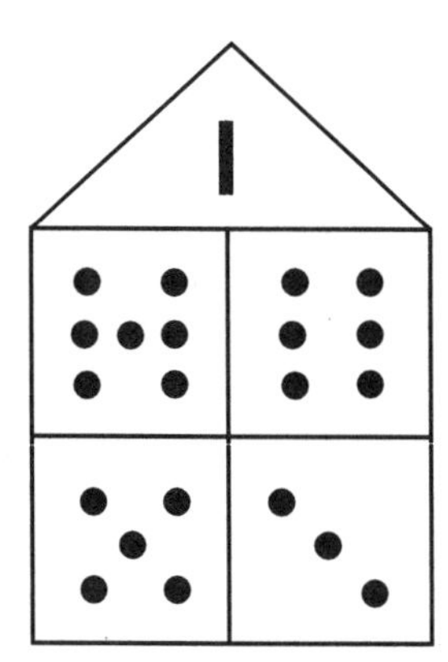

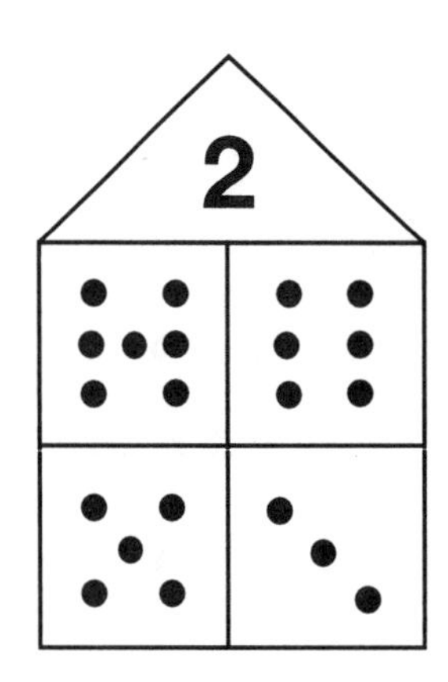

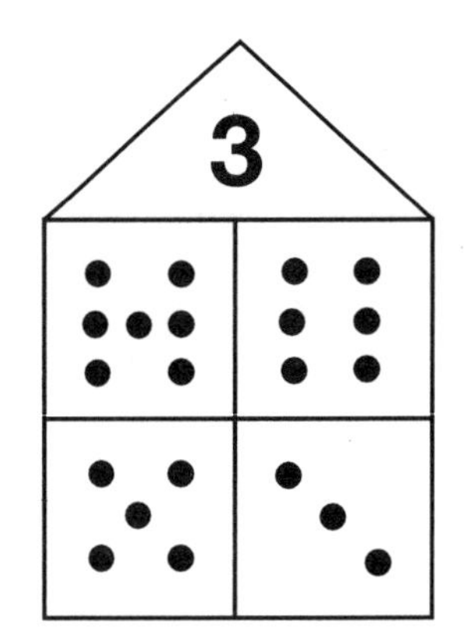

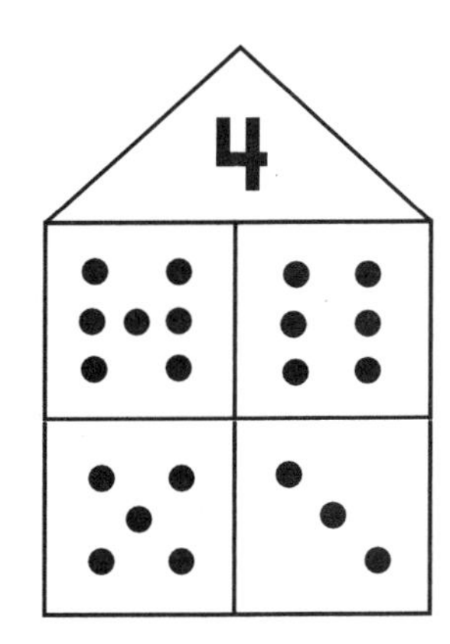

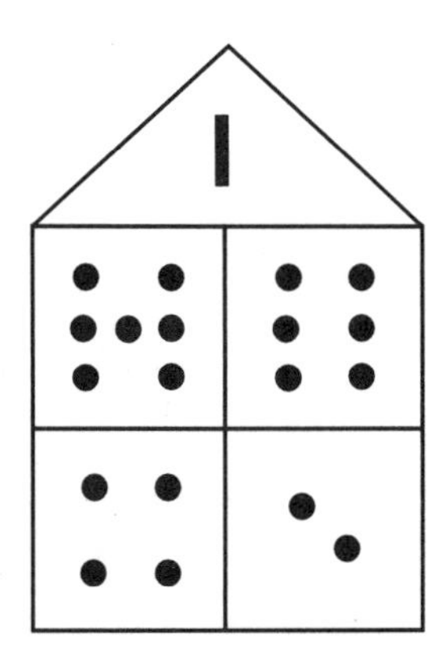

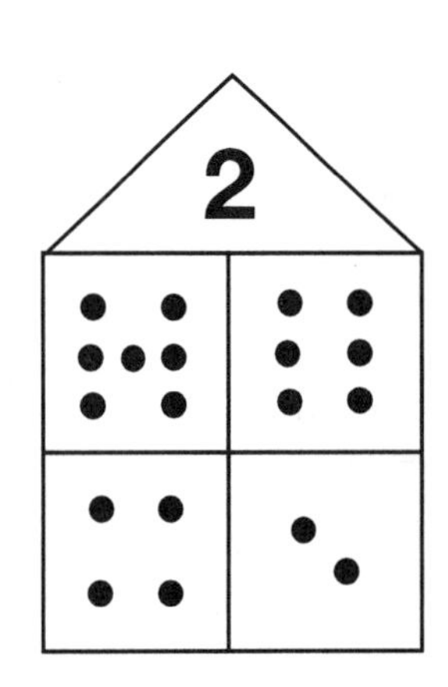

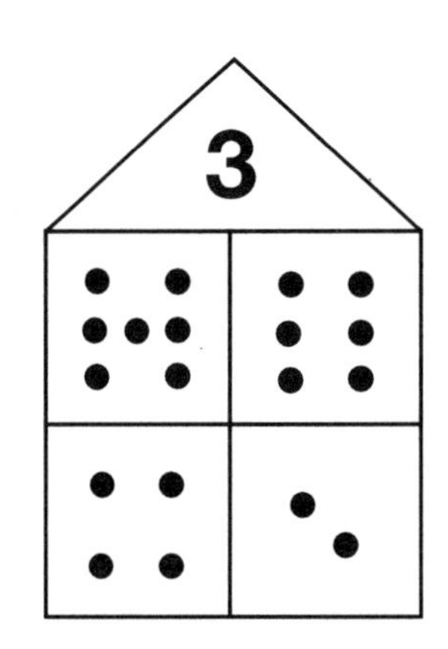

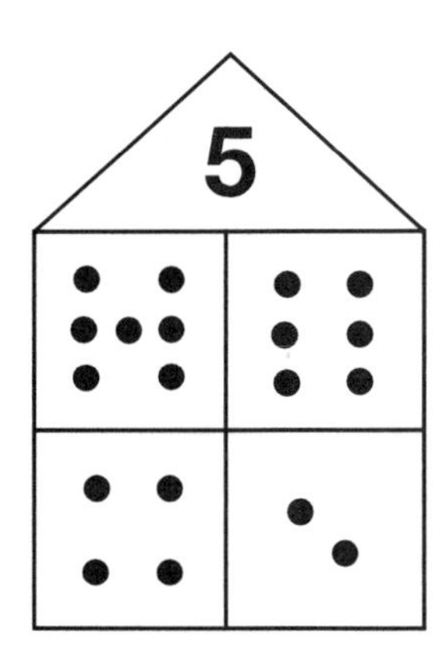

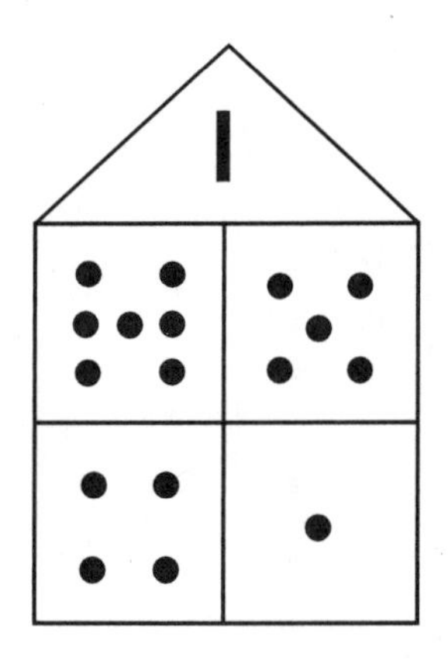

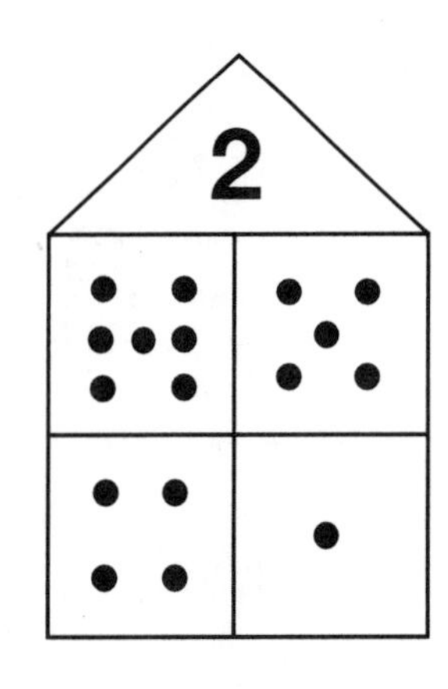

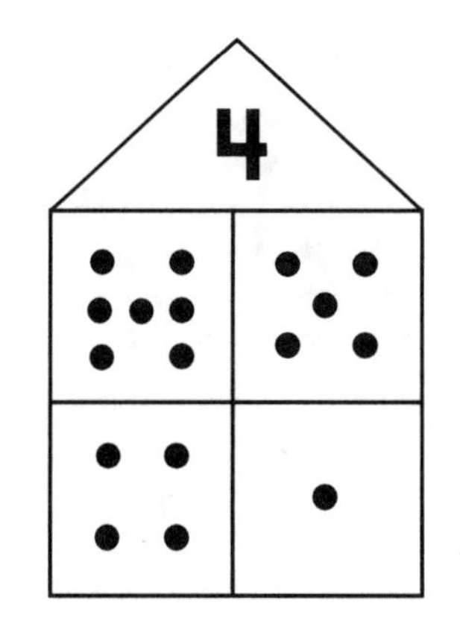

6

Date ____________________ Name ______________________________

Roll and Fill Differences I

Roll a number cube to fill in at least one number for each problem. Will you roll both numbers of some equations?

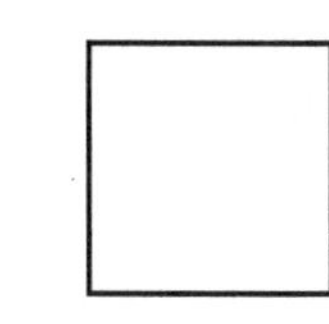
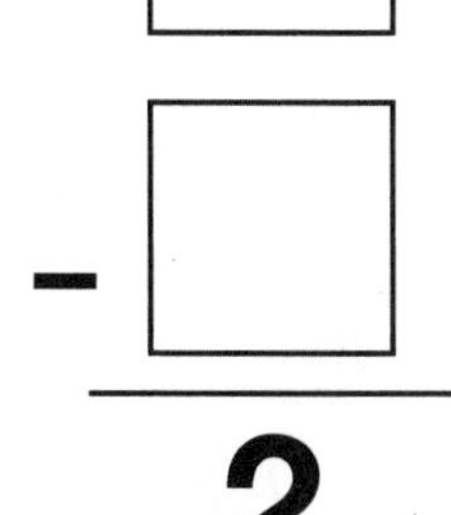
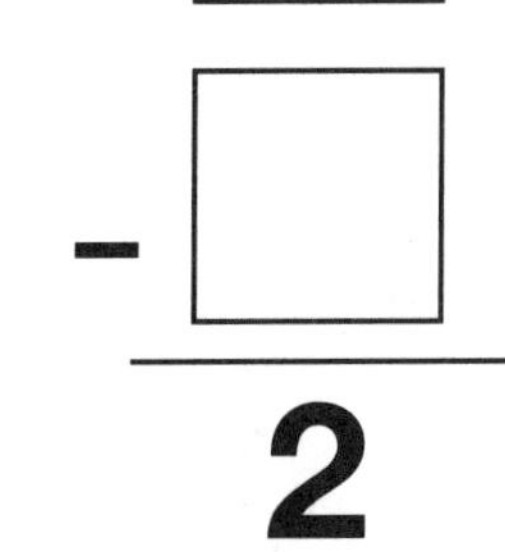

$\square - \square = 1$ $\square - \square = 2$ $\square - \square = 1$ $\square - \square = 2$

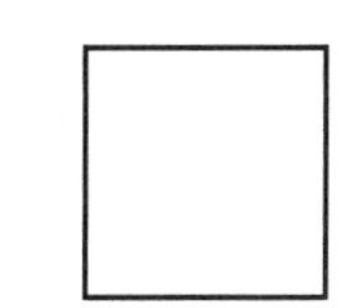

$\square - \square = 0$ $\square - \square = 1$ $\square - \square = 2$ $\square - \square = 3$

$\square - \square = 0$ $\square - \square = 1$

$\square - \square = 2$ $\square - \square = 3$

If time runs out, fill in any numbers to correctly complete the equations.

Date ____________ Name ____________________

Roll and Fill Differences II

Roll a number cube to fill in at least one number for each problem. Will you roll both numbers of some equations?

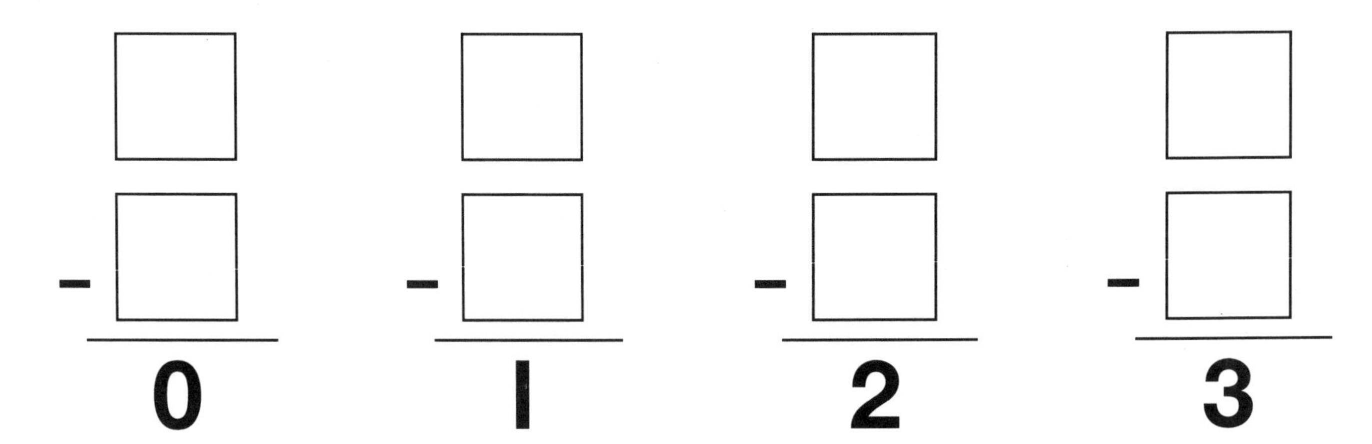

$\square - \square = 4$ $\square - \square = 5$

Roll a number cube and fill in at least two numbers for each problem. How many third numbers can you roll?

$10 - \square - \square = 1$

$11 - \square - \square = 3$

$12 - \square - \square = 5$

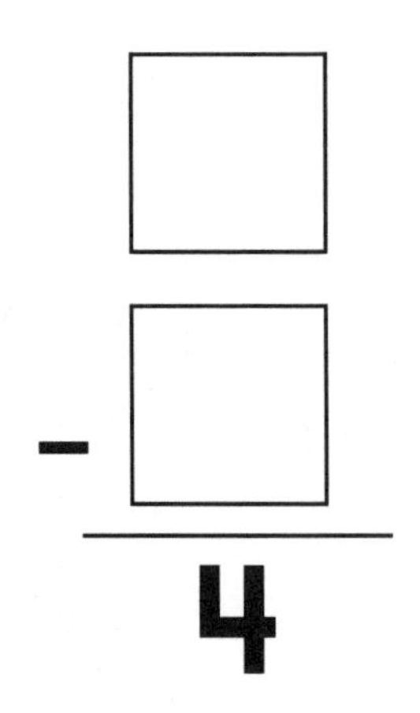

Date ____________ Name ____________________

Equation Hunt I

Use three of these four numbers to make a true subtraction equation.

1.

1	4
~~6~~	3

4 − 3 = 1

2.

2	4
1	1

3.

4	2
0	4

4.

4	2
7	2

5.

8	1
0	1

6.

3	2
6	1

7.

2	5
4	3

8.

8	0
4	8

9.

6	3
4	2

10.

6	6
0	5

11.

6	3
4	3

12.

4	2
7	5

Date ____________ Name ____________________

Equation Hunt II

Use three of these four numbers to make a true subtraction equation.

1.

6 − 5 = 1

2.

4	2
8	4

3.

5	7
1	4

4.

7	5
4	3

5.

7	1
6	4

6.

3	8
1	5

7.

8	5
7	1

8.

3	2
9	6

9.

8	6
2	3

10.

9	3
7	2

11.

9	1
8	2

12.

4	5
9	6

Independent Activity

Mixed Facts

Assumptions The addition and subtraction facts have previously been taught and reviewed, emphasizing understanding. Concrete objects and visual models, such as counters, Ten Frames, and dominoes, have been used extensively. An effort has been made to connect the addition and subtraction facts.

Section Overview and Suggestions

Sponges

How Many Now? p. 82

What's Your Difference? pp. 83–84

What Works? pp. 85–86

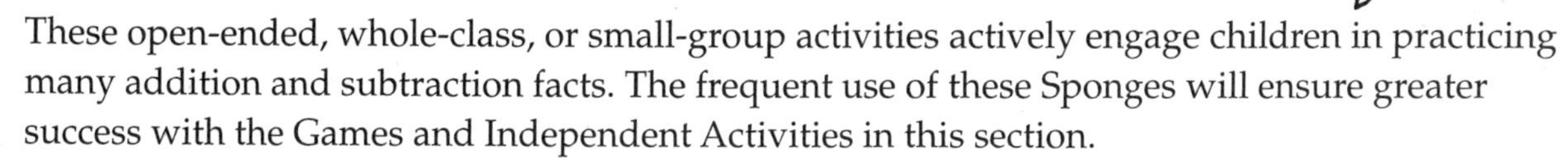

These open-ended, whole-class, or small-group activities actively engage children in practicing many addition and subtraction facts. The frequent use of these Sponges will ensure greater success with the Games and Independent Activities in this section.

Skill Checks

Mixed Facts 1–6 pp. 87–89

These provide a way to help parents, children, and you see children's improvement with the facts. Remember to have children respond to STOP before they solve the ten problems.

Games

Cover Ten pp. 90–91

Add or Subtract pp. 92–94

Pair Search pp. 95–96

These open-ended, repeatable Games actively involve children in practicing many addition and subtraction facts. Each Game requires children to verbalize effective strategies by asking them to discuss potential plays with their partners. In *Cover Ten* and *Pair Search* pairs collaborate to meet the Game objectives. The use of a Dot Cube with a number strip in *Cover Ten* accommodates the counting-on approach. *Add or Subtract* includes two versions; the more challenging one extends to digits four through nine. Repeated experience with *Pair Search* should lead to success on *Seeking Equations,* p. 100.

Independent Activities

Sums and Differences p. 97

What's the Sign? pp. 98–99

Seeking Equations p. 100

These Independent Activities provide long-term practice of varied facts. *Sums and Differences* and *Seeking Equations* are open-ended, repeatable Activities that promote practice of many addition and subtraction facts. *What's the Sign?* includes two activity sheets, allowing students to practice the easier and the more difficult facts.

How Many Now?

Tips If preferred, children might indicate visualized amounts by displaying Digit Cards. If children seem ready for a challenge, have them try to visualize three changes to a displayed amount.

Topic: Addition and Subtraction Facts

Object: Visualize adding and subtracting to displayed amounts.

Groups: Whole class or small group

Materials

- transparency of Ten Frame form, p. 149
- counters

Directions

1. The leader places four to six counters in the displayed Ten Frame and asks children, "How many?"
2. The leader asks children to imagine that two more counters are added to the Ten Frame. The children are asked to show with their fingers the new amount that would be displayed.

   Example: If five is displayed, children visualize a Ten Frame with seven counters, and indicate seven with their fingers.
3. The leader adds two more counters to display seven, and asks children to now imagine one less counter. The children are asked to show with their fingers the new amount that would be displayed.
4. The leader begins again by displaying an amount between three and eight and asks children, "How many?"
5. The leader now asks children to imagine two changes to the displayed amount.

   Example: If six counters are displayed, children are asked to imagine three more and two less. After allowing time for all children to think, children show with their fingers the new amount.

●	●	●	●	●

6. Volunteers help leader display on the Ten Frame how the original amount changed.
7. The leader repeats steps 3–5 with varying amounts. Sometimes the leader asks the children to imagine subtracting amounts first.

Making Connections

Promote reflection and make mathematical connections by asking:

- How did the Ten Frame help you identify the final amount?
- Which amounts are easier to visualize? Why?

What's Your Difference?

Topic: Addition and Subtraction Facts

Object: Identify differences of two numbers when given the sum.

Groups: Whole class or small group

Materials

- transparency of *What's Your Difference?* activity form, p. 84
- Digit Squares for each child, p. 143
- transparent Digit Squares for leader
- 5-by-8-inch card

***Tips** Encourage more subtraction practice by having children first identify differences and change the game to* What's the Sum? *As children gain confidence, have them create* What's Your Difference? *puzzles and help lead the warm-up.*

Directions

1. The leader displays the addition equation on the *What's Your Difference?* activity form and states, "I'm thinking of two numbers. When I add these numbers, I get five. Use your Digit Squares to display two numbers that equal five." As children volunteer their two addends, the leader places corresponding Digit Squares on the addition equation, children read the equation, and the leader moves the pair of numbers to the edge of the overhead.

   Example: For the sum of 5, children would identify 1 and 4, 2 and 3, 5 and 0.

   $\boxed{1} + \boxed{4} = 5$

2. The leader removes the 5-by-8-inch card to display the subtraction equation on the *What's Your Difference?* activity form and asks, "What is the difference of your two numbers? If you subtract the smaller number from the larger number, what is your answer?" As children share possible differences, the leader asks, "Who has the same difference? What are the two numbers?"

   Example: A child responds, "I have a difference of four." Children with 5 and 1 display Digit Squares.

3. As each child shares a new possible difference, the leader asks a similar set of questions.
4. After all possible differences are explored, leader provides a new sum and repeats the sequence.

Making Connections

Promote reflection and make mathematical connections by asking:

- What could we use to help us identify the numbers more quickly?.

What's Your Difference?

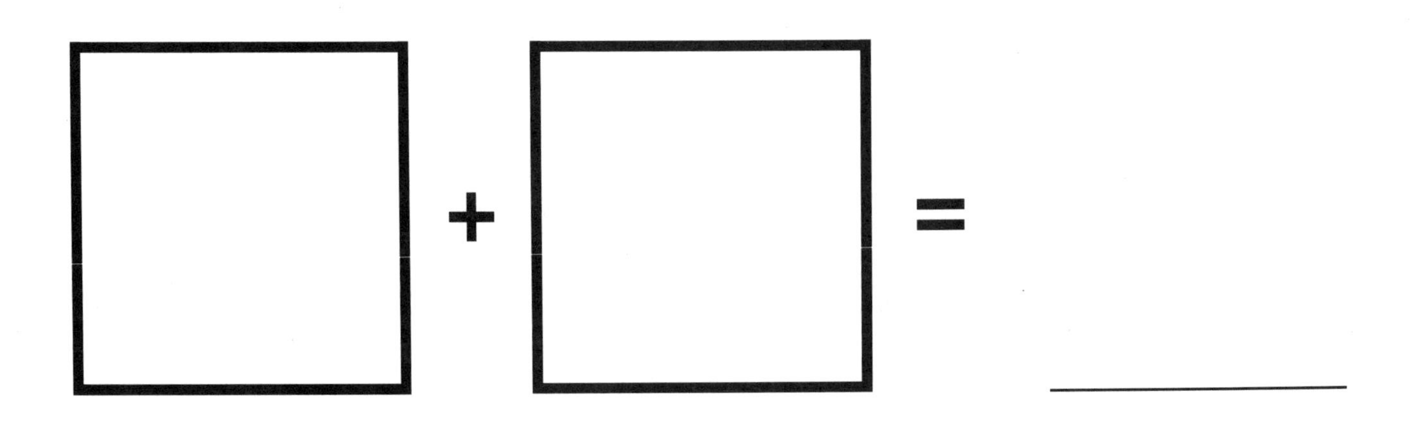

What Works?

Topic: Addition and Subtraction Facts

Object: Select Dot Cards to equal target sums and differences.

Groups: Whole class or small group

Materials

- transparent Dot Pattern Cards, p. 145
- transparency of *What Works?* activity form, p. 86

Directions

1. The leader displays two transparent Dot Pattern Cards, like 2 and 5, on the *What Works?* activity form.
2. The leader states, "By using each of these numbers once, I can make three. Signal if you know the operation I used." Pause to allow thinking time. (Children use fingers to signal subtraction.) Throughout, encourage children to verbalize their equation.
3. The leader continues, "By using each of these numbers once, I can make seven." Pause. "Please signal if you know what operation I used." (Children signal addition.)
4. The leader selects two new Dot Pattern Cards, displays them on the *What Works?* activity form, and repeats steps 1–3.
5. When children seem ready, the leader selects and displays three Dot Pattern Cards. The leader then challenges children to use any two of the three displayed amounts to make specified sums and differences. After children give a solution, they are encouraged to seek additional solutions for the same amount.

   Example: Display 1, 2, and 5. Say, "By using any two of these numbers, how could you make 4?" Pause. "If you have found a way, signal the operation you used. Is there another way?"
6. The leader continues to display the same three Dot Pattern Cards and asks children how they could make new target amounts. After children verbalize their equations, the leader asks if there are other ways to equal the target amount.
7. The leader repeats this process with a new combination of three Dot Pattern Cards.

Tips *When children seem ready, transition into displaying three Digit Cards. Reinforce symbolic representation by having children write the equation for each found sum or difference.*

Making Connections

Promote reflection and make mathematical connections by asking:

- Which target numbers are easier to make? Please explain.
- How did using three numbers change this activity?

What Works?

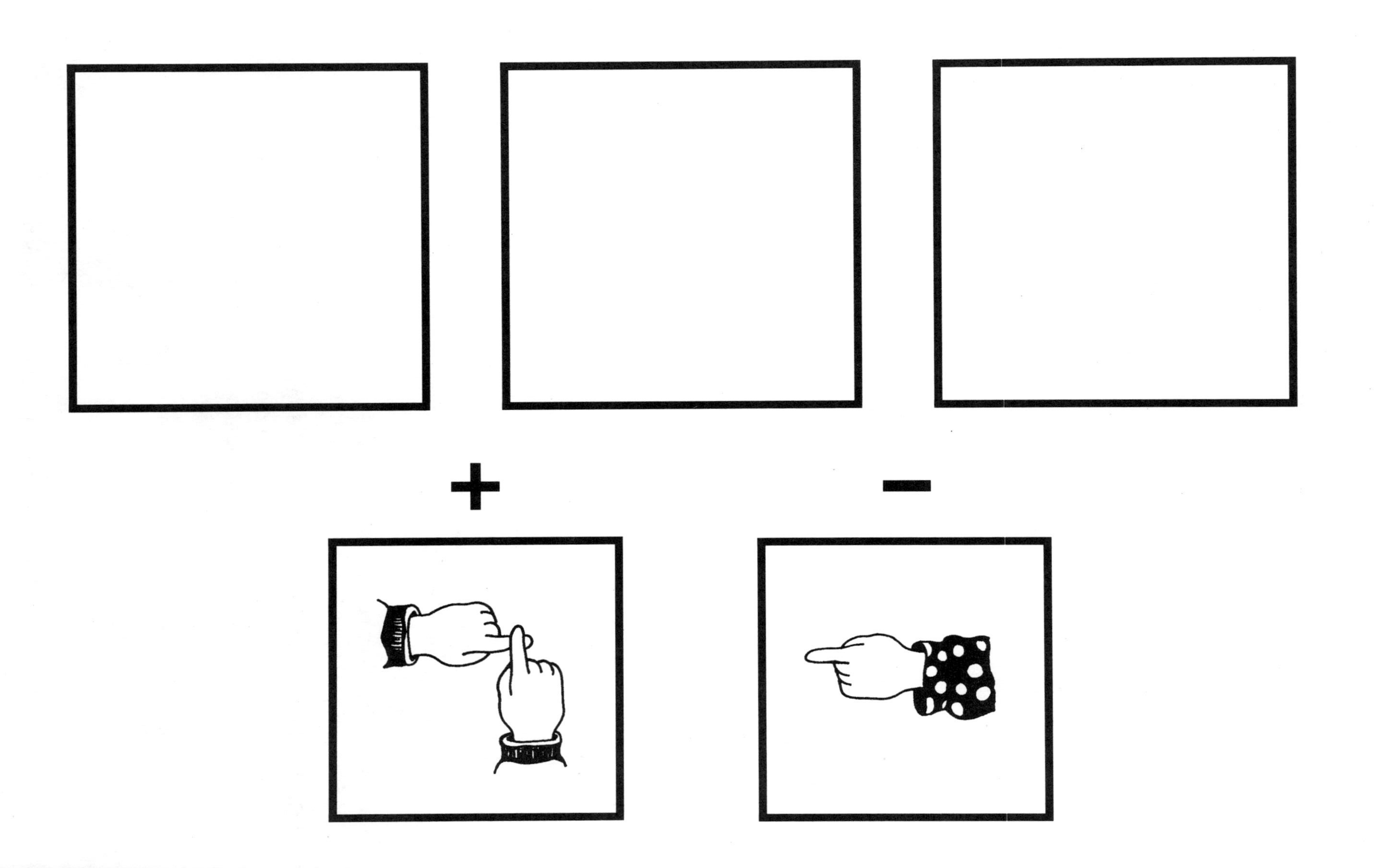

Date ______________ Name ______________________

Mixed Facts 1

Don't start yet! Circle two problems that may have answers greater than 8.

1. $4 - 1 =$ ______ **2.** $7 + 2 =$ ______

3. $\begin{array}{r} 3 \\ +\ 4 \\ \hline \end{array}$ **4.** $\begin{array}{r} 12 \\ -\ 6 \\ \hline \end{array}$ **5.** $\begin{array}{r} 8 \\ +\ 4 \\ \hline \end{array}$ **6.** $\begin{array}{r} 14 \\ -\ 6 \\ \hline \end{array}$

Fill in the sign. **7.** $6 \triangle 2 = 4$ **8.** $5 \triangle 4 = 9$

9. $10 - 2 + 3 =$ ______ **10.** $3 + 6 +$ ______ $= 11$

Write two different equations to equal 7.
Include subtraction.

Date ______________ Name ______________________

Mixed Facts 2

STOP Don't start yet! Circle two problems that may have odd answers.

1. $5 - 2 =$ ______ **2.** $8 + 1 =$ ______

3. $\begin{array}{r} 6 \\ +\ 3 \\ \hline \end{array}$ **4.** $\begin{array}{r} 9 \\ -\ 7 \\ \hline \end{array}$ **5.** $\begin{array}{r} 9 \\ +\ 4 \\ \hline \end{array}$ **6.** $\begin{array}{r} 13 \\ -\ 9 \\ \hline \end{array}$

Fill in the sign. **7.** $6 \triangle 2 = 8$ **8.** $8 \triangle 3 = 5$

9. $7 + 3 - 2 =$ ______ **10.** $6 + 5 +$ ______ $= 12$

Go On Use these numbers to come up with two different answers: 12, 3

Date ________________ Name ____________________________

Mixed Facts 3

Don't start yet! Circle two problems that may have answers greater than 10.

1. 6 – 1 = ______ **2.** 9 + 2 = ______

3. 5 + 4 = ___ **4.** 11 – 6 = ___ **5.** 7 + 6 = ___ **6.** 15 – 8 = ___

Fill in the sign. **7.** 7 △ 2 = 9 **8.** 7 △ 5 = 2

9. 9 – 2 + 1 = ______ **10.** 5 + 4 + ______ = 10

What numbers are missing? 2, 5, ____ , 11, ____ , ____
What is the rule?

Date ________________ Name ____________________________

Mixed Facts 4

Don't start yet! Circle two problems that may have an even answer.

1. 7 – 2 = ______ **2.** 10 + 1 = ______

3. 4 + 2 = ___ **4.** 10 – 7 = ___ **5.** 8 + 6 = ___ **6.** 14 – 9 = ___

Fill in the sign. **7.** 4 △ 1 = 3 **8.** 3 △ 4 = 7

9. 6 + 4 – 3 = ______ **10.** 6 + 3 + ______ = 11

Go On Make two different equations to equal 11. Include subtraction.

Date ________________ Name ____________________

Mixed Facts 5

Don't start yet! Circle two problems that may have answers less than 10.

1. 9 − 1 = ______

2. 11 + 2 = ______

3. 5 + 3 = ___

4. 11 − 7 = ___

5. 9 + 5 = ___

6. 15 − 6 = ___

Fill in the sign. **7.** 7 △ 3 = 4

8. 6 △ 3 = 9

9. 8 − 2 + 3 = ______

10. 4 + 4 + ______ = 10

Go On Use these numbers to come up with two different answers: 9, 6

Date ________________ Name ____________________

Mixed Facts 6

STOP Don't start yet! Circle two problems that may have answers less than 8.

1. 8 − 2 = ______

2. 12 + 1 = ______

3. 6 + 4 = ___

4. 12 − 5 = ___

5. 7 + 5 = ___

6. 13 − 9 = ___

Fill in the sign. **7.** 5 △ 2 = 7

8. 9 △ 7 = 2

9. 5 + 4 − 2 = ______

10. 7 + 4 + ______ = 12

Go On What numbers are missing? 17, 15, ____, ____, 9, 7, ____, ____, 1
What is the rule?

Cover Ten

Topic: Addition and Subtraction Facts

Object: Cover ten numbers.

Groups: Pairs

Materials for each group

- *Cover Ten* gameboard, p. 91
- Dot Cube (1–6), p. 144
- 10 counters

Tip By replacing the Dot Cube with a Number Cube, more children will be ready to try the more challenging games in this section.

Directions

1. In this game, a pair works cooperatively to cover numbers.
2. The pair rolls the Dot Cube and places it near the 6–9 number strip.
3. The pair adds or subtracts the rolled amount with any number on the number strip. The pair covers the resulting sum or difference with a counter.

   Example: If 3 is rolled, the pair may choose to subtract 3 from 8 and cover 5.
4. Each turn the pair discusses the various options in order to agree on where to place the counter.
5. Each member of the pair alternates turns, rolling the Dot Cube and covering the resulting amounts.
6. When the pair covers ten numbers with counters, the game ends.

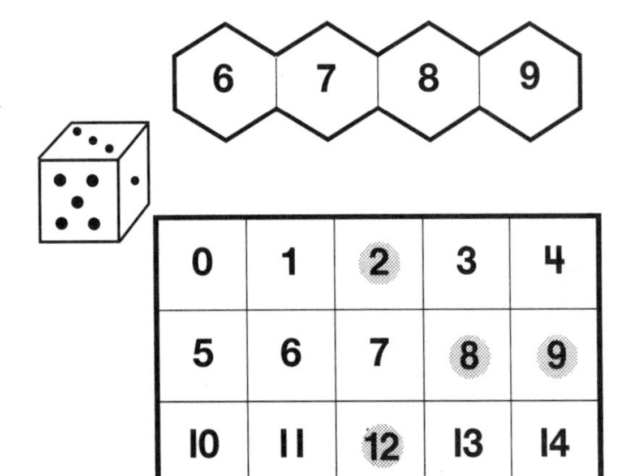

Making Connections

Promote reflection and make mathematical connections by asking:

- Which amounts were easy to make?

Cover Ten

0	1	2	3	4
5	6	7	8	9
10	11	12	13	14

Add or Subtract

Tip *If the game is played at home, require adult players to get four-in-a-row.*

Topic: Addition and Subtraction Facts

Object: Cover three numbers in a row with your counters.

Groups: 2 pair players

Materials for each group

- *Add or Subtract A* gameboard, p. 93
- 2 Number Cubes (1–6), p. 147
- counters (different kind for each pair)

Directions

1. The first pair rolls two Number Cubes. The pair decides whether to add or subtract the displayed amounts.
2. Next the pair states the equation and places a counter on the resulting sum or difference.

   Example: If 2 and 5 are rolled, the pair might cover 3 (5 – 2) or 7 (2 + 5).
3. If a pair rolls two sixes, the pair is allowed to roll again.
4. Pairs alternate turns rolling Number Cubes, stating equations, and placing counters on the gameboard.
5. The first pair to have three counters in a row horizontally, vertically, or diagonally wins.

   Note: Using the *Add or Subtract B* gameboard with only one regular Number Cube (1–6) and a special Number Cube (4–9), p. 147, challenges children to use higher numbers.

7	5	3	10
2	9	6	1
6	4	2	7
1	3	8	0
4	1	5	2

Making Connections

Promote reflection and make mathematical connections by asking:

- What strategies helped you line up your counters in a row?

Add or Subtract A

7	5	3	10
2	9	6	1
6	4	2	7
1	3	8	0
4	1	5	2

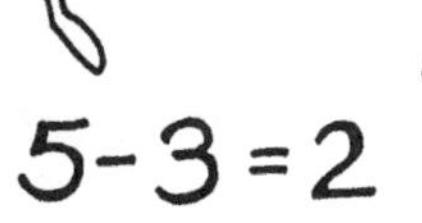

Add or Subtract B

5	2	3	9
12	7	8	11
5	1	3	6
0	6	2	8
4	10	7	1

$7 + 3 = 10$

$7 - 3 = 4$

Pair Search

Topic: Addition and Subtraction Facts

Object: Make equations to ten.

Groups: Pairs

Tips *Some children will benefit if dot cubes replace Number Cubes. If desired, the gameboard can be duplicated and used as a recording sheet and a possible homework option.*

Materials for each group

- *Pair Search* gameboard, p. 96
- 3 Number Cubes (1–6), p. 147
- counters

Directions

1. In this game, a pair works cooperatively to complete each equation on the gameboard.
2. One member of the pair rolls the three Number Cubes.
3. The pair selects two of the three Number Cubes to add or subtract, temporarily places the two selected cubes on the gameboard to display an equation, states the equation, and covers the resulting sum or difference with a counter.

   Example: If 2, 3, and 4 are rolled, the pair might display and cover 1 (3 – 2), 2 (4 – 2), 5 (2 + 3), 6 (2 + 4), or 7 (3 + 4).

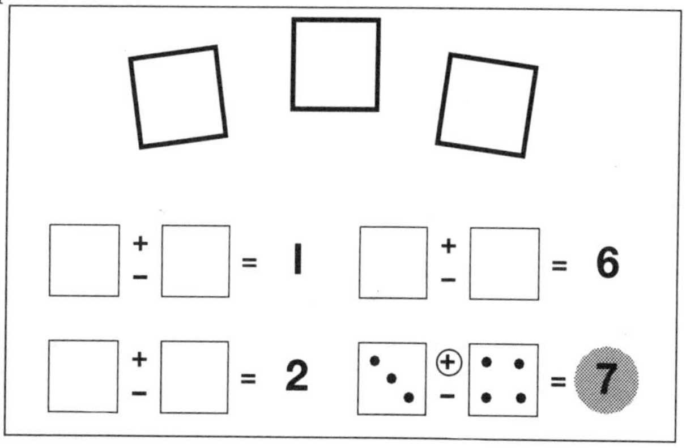

4. Each member of the pair alternates turns rolling and placing Number Cubes, stating equations, and placing counters on the gameboard. During each turn, the pair discusses options and agrees on selected equations.
5. If all the possible resulting sums or differences are already covered, the pair rolls again.
6. The pair repeats these steps until they have found equations for each of the ten amounts shown on the gameboard.

Making Connections

Promote reflection and make mathematical connections by asking:

- Which numbers were easier to make? Why?

Pair Search

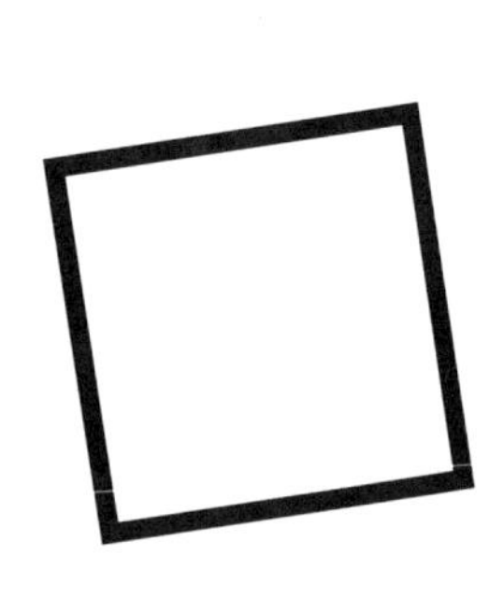

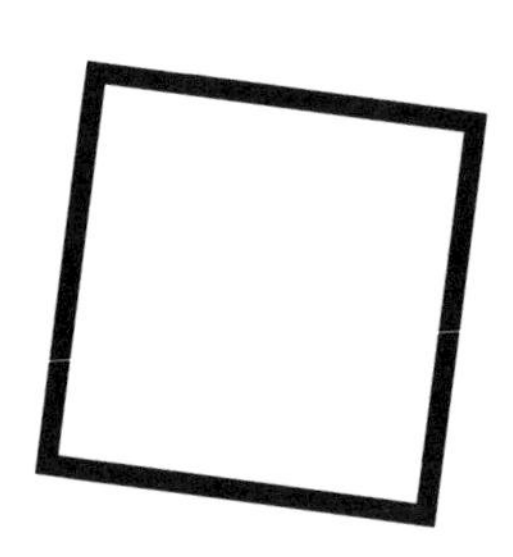

☐ ± ☐ = 1 ☐ ± ☐ = 6

☐ ± ☐ = 2 ☐ ± ☐ = 7

☐ ± ☐ = 3 ☐ ± ☐ = 8

☐ ± ☐ = 4 ☐ ± ☐ = 9

☐ ± ☐ = 5 ☐ ± ☐ = 10

Date ________________ Name ____________________________

Sums and Differences

Roll two number cubes. After you record the numbers rolled, record the sum and the difference for your two numbers.*

My Roll		My Sum	My Difference
______	______	____________	____________
______	______	____________	____________
______	______	____________	____________
______	______	____________	____________
______	______	____________	____________
______	______	____________	____________
______	______	____________	____________
______	______	____________	____________
______	______	____________	____________
______	______	____________	____________

* If students need additional practice writing equations, have them write the whole equation.

Date ____________ Name ____________________

What's the Sign? I

Write a plus or minus in each triangle to make correct equations.

1.
$$\begin{array}{r} 3 \\ \triangle\ 1 \\ \hline 2 \end{array}$$

2.
$$\begin{array}{r} 6 \\ \triangle\ 3 \\ \hline 9 \end{array}$$

3.
$$\begin{array}{r} 4 \\ \triangle\ 3 \\ \hline 7 \end{array}$$

4.
$$\begin{array}{r} 5 \\ \triangle\ 2 \\ \hline 3 \end{array}$$

5.
$$\begin{array}{r} 6 \\ \triangle\ 6 \\ \hline 0 \end{array}$$

6.
$$\begin{array}{r} 9 \\ \triangle\ 5 \\ \hline 4 \end{array}$$

7.
$$\begin{array}{r} 3 \\ \triangle\ 5 \\ \hline 8 \end{array}$$

8.
$$\begin{array}{r} 7 \\ \triangle\ 2 \\ \hline 9 \end{array}$$

9. $6 \triangle 2 = 8$

10. $7 \triangle 2 = 5$

11. $5 \triangle 4 = 1$

12. $8 \triangle 4 = 4$

13. $9 \triangle 1 = 8$

14. $5 \triangle 4 = 9$

Date ______________ Name ______________________

What's the Sign? II

Write a plus or minus in each triangle to make correct equations.

1. $6 \triangle 4 = 10$

2. $12 \triangle 6 = 6$

3. $3 \triangle 8 = 11$

4. $13 \triangle 4 = 9$

5. $11 \triangle 2 = 9$

6. $12 \triangle 7 = 5$

7. $7 \triangle 6 = 13$

8. $12 \triangle 8 = 4$

9. $9 \triangle 3 = 12$

10. $10 \triangle 7 = 3$

11. $11 \triangle 6 = 5$

12. $7 \triangle 4 = 11$

13. $13 \triangle 5 = 8$

14. $11 \triangle 3 = 8$

Date ____________________ Name ________________________________

Seeking Equations

From a set of 1–6 Digit Squares, place four Digit Squares below.

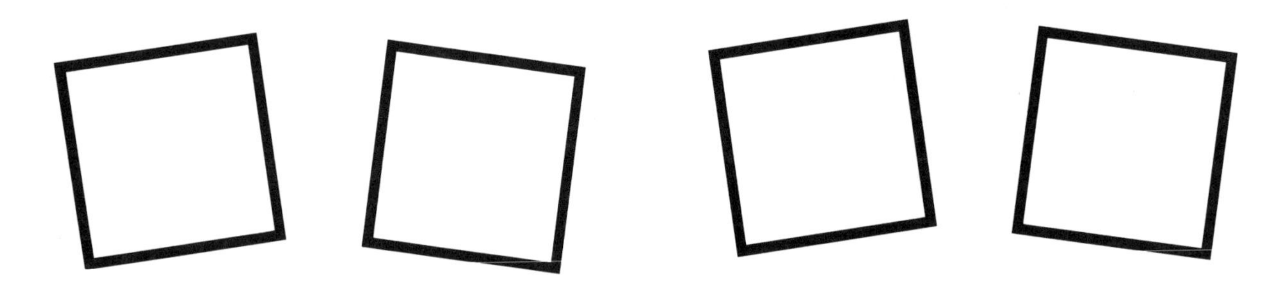

For each equation below, use only two of the four Digit Squares. Record the numbers and circle the plus or minus sign to make each equation correct.

How many equations can you make?

☐ $\pm$ ☐ = 1

☐ $\pm$ ☐ = 2

☐ $\pm$ ☐ = 3

☐ $\pm$ ☐ = 4

☐ $\pm$ ☐ = 5

☐ $\pm$ ☐ = 6

☐ $\pm$ ☐ = 7

☐ $\pm$ ☐ = 8

☐ $\pm$ ☐ = 9

☐ $\pm$ ☐ = 10

Place Value

Assumptions Place value concepts have been taught for understanding with an emphasis on developing number sense. Children have counted, grouped, and traded concrete objects to form tens and ones. Concrete and visual models such as Ten Frames and 100 Charts have been used extensively.

Section Overview and Suggestions

Sponges

Name the Amount pp. 102–103

Show Me pp. 104–105

50 Chart Pieces pp. 106–107

Find My Number pp. 108–109

All the Sponges help children develop memory of the 50 Chart or 100 Chart, a valuable visual tool. Eventually children might solve these warm-ups mentally without the use of the 50 Chart. Continued use of *Show Me* ensures success with other Sponges and Games, and provides opportunities for children to see multiple representations of numbers. *Find My Number* helps children see number relationships on the 50 Chart.

Skill Checks

What's in That Place? 1–6 pp. 110–112

The Skill Checks provide a way to help parents, children, and you see a child's improvement with place value. Remember to have all children respond to STOP before solving the ten problems.

Games

Race to 40 pp. 113–114

Claim All You Can pp. 115–116

Low-High Spin pp. 117–118

These open-ended, repeatable Games further the development of number sense and understanding of number relationships. To ensure success with these thinking Games, allow repeated experience with the Sponges, especially *Find My Number* which leads into *Claim All You Can. Race to 40* asks children to group and trade objects by tens and ones. *Low-High Spin,* a cooperative Game, promotes strategic thinking as children decide where to place two-digit numbers.

Independent Activities

What Numbers Are Missing? pp. 119–120

Fill in the Pieces pp. 121–122

Smallest and Largest pp. 123–124

What Numbers Are Missing? and *Fill in the Pieces* engage children in visualizing numbers on the 50 or 100 Chart. Children will benefit by using Digit Squares to solve *Smallest and Largest,* as they create and order two-digit numbers. Each Independent Activity has two levels of difficulty and builds number sense as children enhance their place value skills.

Name the Amount

Topic: Number Sense to 100

Object: Recognize visual representations of numbers to 100.

Groups: Whole class or small group

Tip *Emphasize numbers under 50 at first and proceed cautiously to numbers beyond.*

Materials

- transparency of *Name the Amount*, p. 103
- two 5-by-8-inch cards
- 100 Chart for each child, p. 139
- counter for each child

Directions

1. The leader makes the desired arrangement with overhead projector off. The leader displays two rows of ten stars (one 5-by-8-inch card covers eight rows of ten stars). The leader asks children to identify the amount shown. Children indicate their responses, by covering that number on their 100 Charts with counters. The leader asks, "How many rows of ten?" (Children respond, "There are two rows of ten.") "How many stars?" (Children respond, "20.")
2. The leader continues this procedure by displaying amounts equaling various multiples of ten, and having children verbalize and identify each displayed amount.
3. Using a second 5-by-8-inch card, the leader displays amounts other than multiples of ten. When showing a partial row, the leader displays stars from left only.

   Example: Leader displays 34 and asks, "How many rows of ten stars and how many extra stars are there?" (Children respond, "Three rows of ten stars and four extra stars.") "How many stars are there?" (Children respond, "34.")

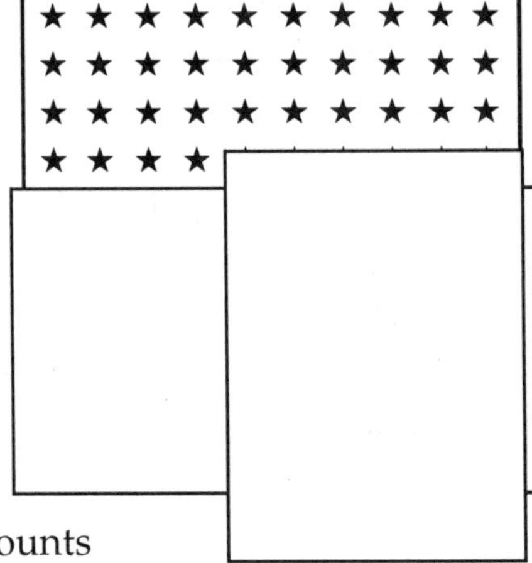

4. The children indicate their response by covering the matching number on their 100 Chart.
5. The leader continues to provide visual clues for various numbers on the 100 Chart and the children identify the amounts with their one counter on their 100 Chart.

Making Connections

Promote reflection and make mathematical connections by asking:

- What helped you easily identify the displayed amounts?
- How was the 100 Chart helpful in this activity?

Name the Amount

Show Me

Topic: Place Value of Tens and Ones

Object: Identify tens and ones and corresponding numbers.

Groups: Whole class or small group

Materials

- Mini Ten Frames sheet cut apart, p. 150 (one sheet supplies 2 children)
- transparency of Mini Ten Frames sheet cut apart
- 20 beans
- 50 Chart for each child, p. 138
- transparency of 50 Chart
- counter for each child
- pencil and 2 paper clips
- transparency of *Show Me* spinners, p. 105

__Tips__ It's best to use no more than two versions of this Sponge during one sitting. This warm-up can easily be extended to a 100 Chart.

Directions

1. The leader spins the two spinners and asks the children to display this amount with their Ten Frames and beans.
2. The leader asks, "What number are you displaying?"
3. The children clear their Ten Frames and the leader spins again.
4. The leader repeats this procedure many times until the children seem ready for a more challenging version.
5. Instead of spinning, the leader displays an amount by using Mini Ten Frames and beans. The children respond by covering the corresponding number on their 50 Charts.
6. Once children become skilled with this version, the leader can vary the abstraction level by choosing one of the following actions:
   - Leader spins an amount and children indicate the number on their 50 Chart.
   - Leader indicates one number on the 50 Chart and children display the amount with Ten Frames and beans.
   - Leader verbalizes an amount, like "three tens and seven," and children indicate the number on their 50 Chart or with Ten Frames and beans.
   - Leader displays a written form of a number, such as two tens and nine, and children respond.

Making Connections

Promote reflection and make mathematical connections by asking:

- Which way do you prefer to display numbers? Why?

Show Me

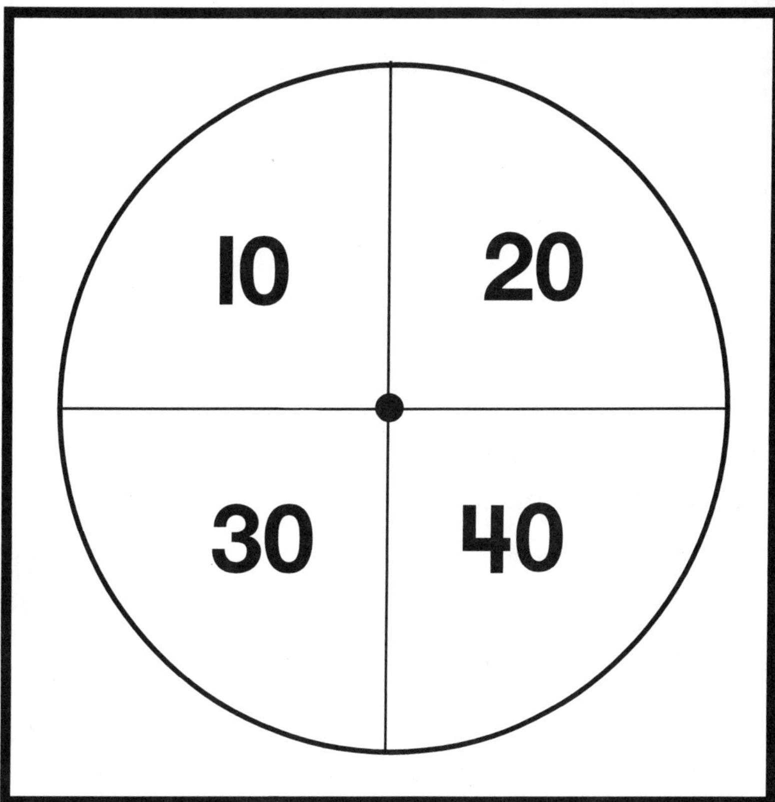

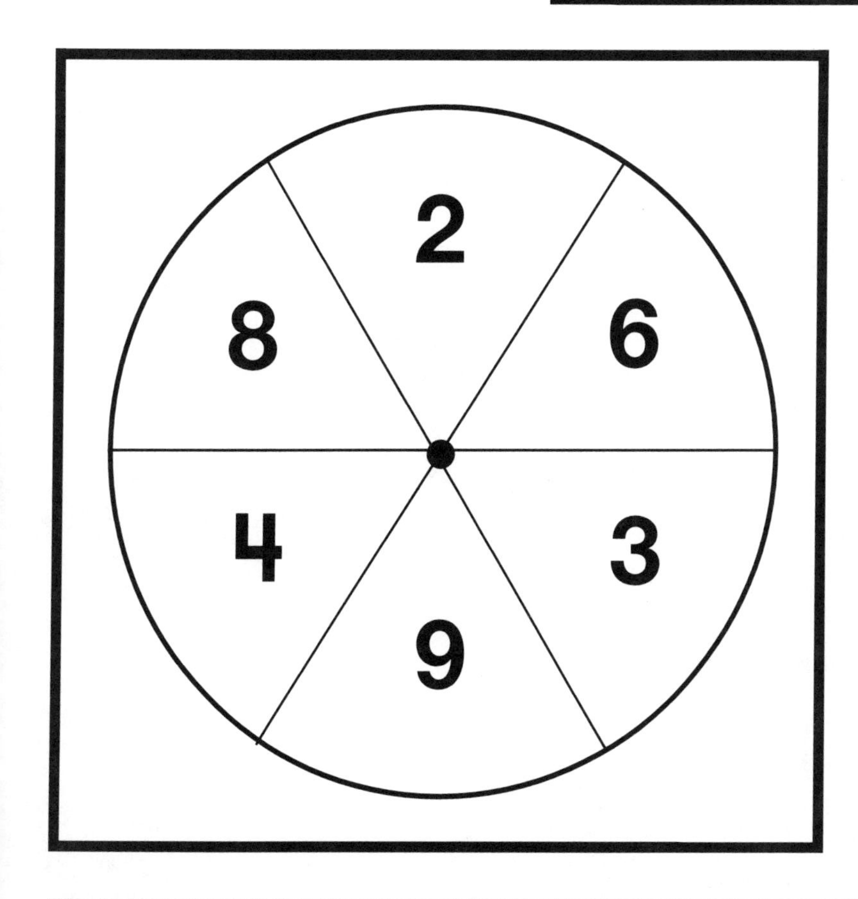

50 Chart Pieces

Tip As children become competent, extend this warm-up by using a 100 Chart.

Topic: Number Relationships

Object: Determine unknown numbers by visualizing the 50 Chart.

Groups: Whole class or small group

Materials

- transparent patterns of *50 Chart Pieces,* p. 107
- 50 Chart for each child, p. 138
- counter for each child

Directions

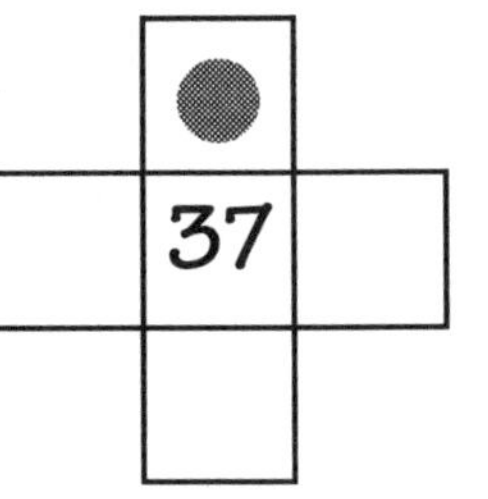

1. The leader displays one of the simpler 50 Chart pieces and writes a number in one of the cells.
2. The leader places a counter in one of the remaining empty cells.
3. Children try to determine what number belongs in the marked empty cell. They indicate their response by covering that number on their 50 Chart.
4. The leader continues to provide more challenging clues, and the players identify the unknown numbers with their one counter on their 50 Chart.
5. As this warm-up proceeds, the leader asks children to explain how they figured out the number. It's helpful to have different approaches shared.
6. Increase the difficulty by having children determine the answer before referring to the 50 Chart.

Making Connections

Promote reflection and make mathematical connections by asking:

- What patterns helped you identify the unknown number?

50 Chart Pieces

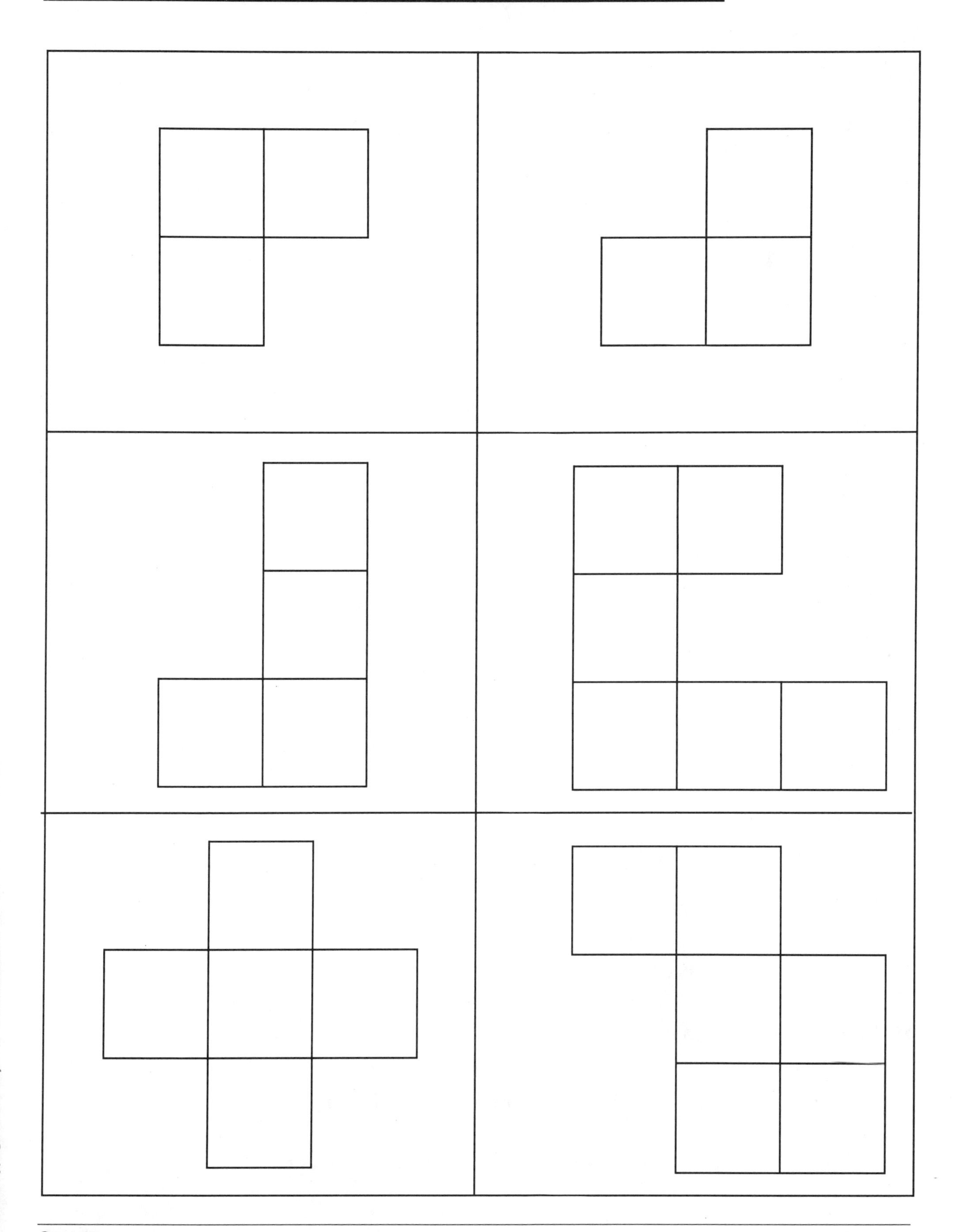

Find My Number

Topic: Number Relationships

Object: Identify related numbers.

Groups: Whole class or small group

Materials

- transparency of *Find My Number* spinner, p. 109
- paper clip and pencil for spinner
- 50 Chart for each child, p. 138
- 2 transparent counters for each child

Tip *As children gain confidence with this warm-up, expand it to include numbers beyond 50 by adding rows of tens, one row at a time.*

Starting Number

Directions

1. The leader displays a Starting Number between 10 and 40.
2. Children individually locate and cover the matching numeral on their 50 Chart with a counter.
3. Using a paper clip and pencil, the leader spins the four-sectioned spinner to let children know how to relate the next number to the starting number. The children use another counter to identify the second number.

   Example: If the starting number is 36 and "10 less" is spun, children indicate 26 with their second counter.
4. The leader asks, "What is your second number? How are your two covered numbers related?"
5. Children remove the second counter. Keeping the same starting number, the leader spins to identify a new number relationship. If the spinner indicates an amount not found on a 50 Chart, the leader spins again.
6. After using a starting number for several rounds, the leader displays a new starting number, spins, and has children identify the resulting numbers and relationships.
7. When children seem confident, the leader repeats the entire procedure with the more challenging six-sectioned spinner.

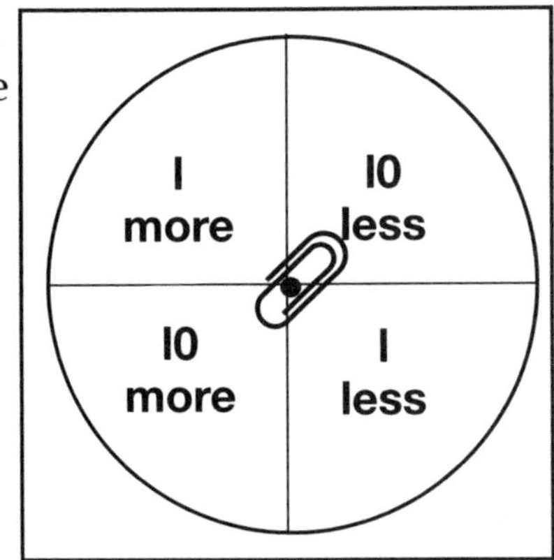

Making Connections

Promote reflection and make mathematical connections by asking:

- If you move up one row on a 50 Chart, how does the number change?
- What patterns did you notice as you identified the related numbers?

Find My Number

Starting Number

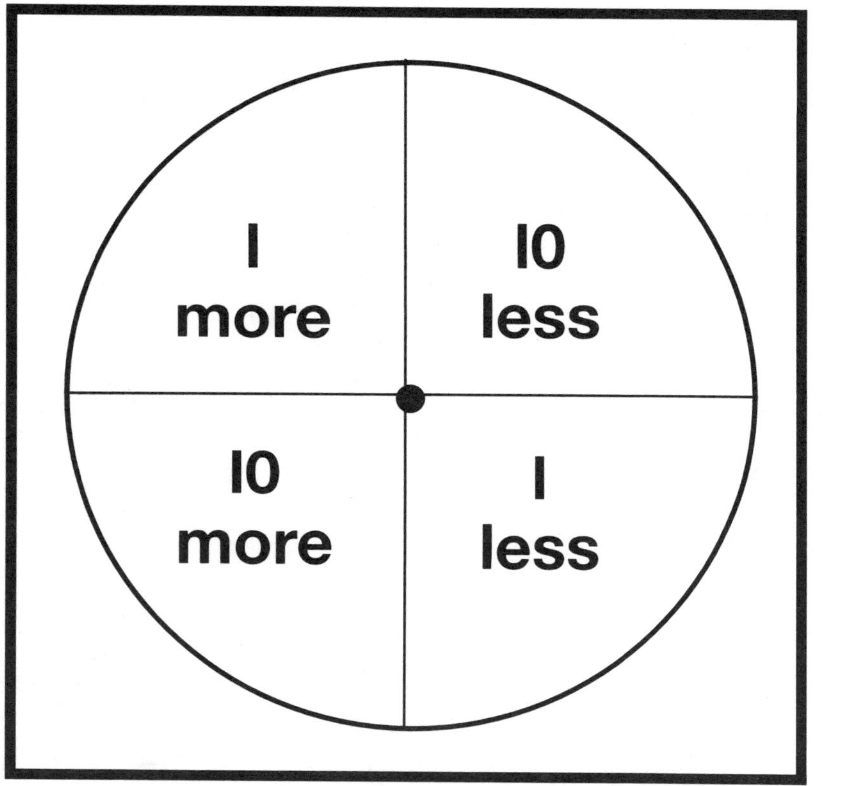

Find My Number

Starting Number

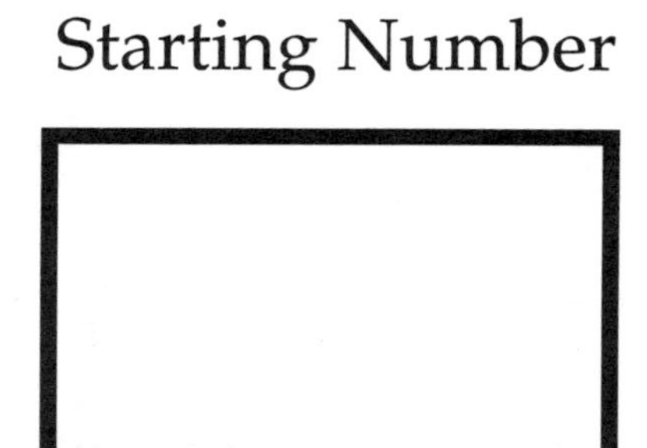

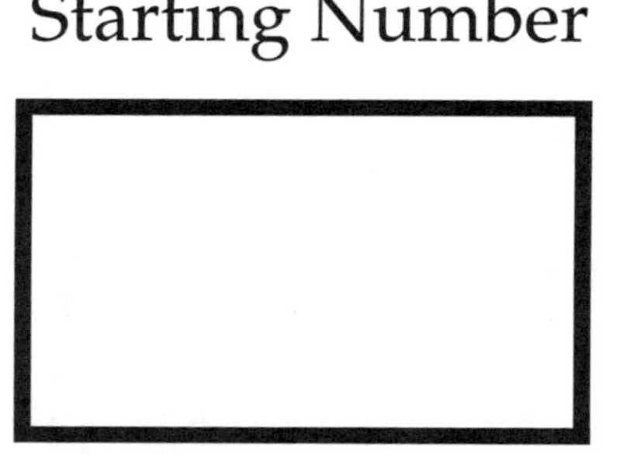

10 less

2 more

20 more

20 less

2 less

10 more

Date ____________ Name ____________

What's in That Place? 1

Don't solve yet! Circle two problems that may have answers less than 40.

Fill in the missing numbers.

1. 24

2. 47

3. one more than 54 ________

4. one less than 73 ________

Order these numbers from smallest to largest.

5. 15, 12, 18 ________

6. 27, 47, 37 ________

7. ten more than 35 ________

8. ten less than 29 ________

9. 40 + 9 = ________
4 + 9 = ________

10. 3 + 7 = ________
30 + 7 = ________

Go On What numbers are missing? 17, 27, 37, ______ , ______ , ______
What is the rule?

Date ____________ Name ____________

What's in That Place? 2

Don't solve yet! Circle two problems that may have answers greater than 60.

Fill in the missing numbers.

1.

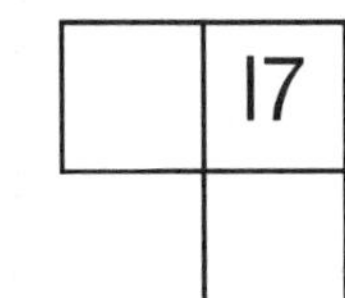

2.

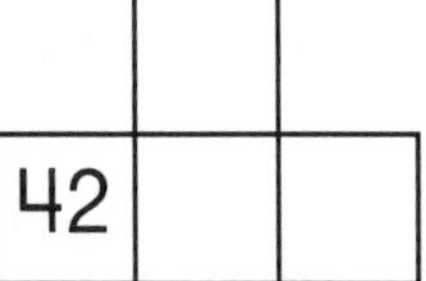

3. one more than 82 ________

4. one less than 68 ________

Order these numbers from smallest to largest.

5. 38, 36, 32 ________

6. 34, 44, 14 ________

7. ten more than 46 ________

8. ten less than 33 ________

9. 2 + 9 = ________
20 + 9 = ________

10. 40 + 7 = ________
4 + 7 = ________

Go On Write three numbers between 25 and 35.
How do you know they fit?

Date ____________ Name ____________________

What's in That Place? 3

STOP Don't solve yet! Circle a problem that may have an answer between 30 and 40.

Fill in the missing numbers.

1. [ ] [ ] / [33]

2. [28] / [ ] [ ] [ ]

3. one more than 67 ______

4. one less than 75 ______

Order these numbers from smallest to largest.

5. 14, 19, 13 ________

6. 56, 26, 46 ________

7. ten more than 24 ________

8. ten less than 41 ________

9.
30 + 8 = ________
3 + 8 = ________

10.
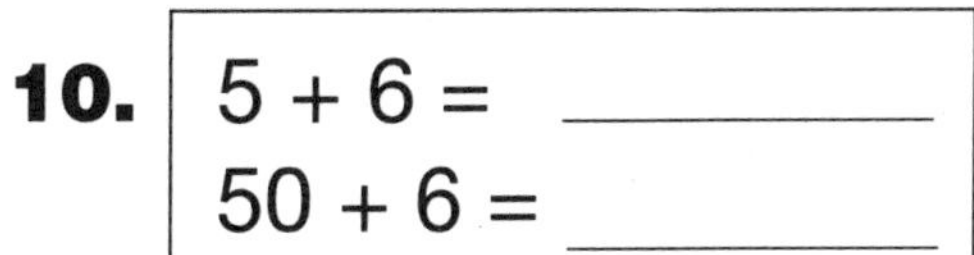
5 + 6 = ________
50 + 6 = ________

Go On Order these numbers from smallest to largest. 52, 83, 38, 61
How do you know?

Date ____________ Name ____________________

What's in That Place? 4

STOP Don't solve yet! Circle a problem that may have an answer between 60 and 80.

Fill in the missing numbers.

1. [26] [ ] / [ ]

2.
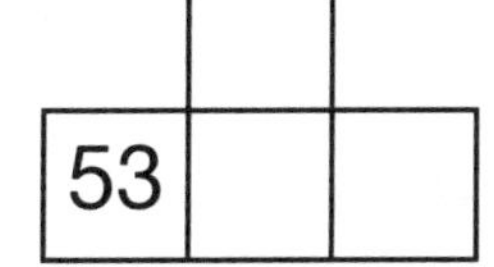
[ ] / [53] [ ] [ ]

3. one more than 94 ________

4. one more than 51 ________

Order these numbers from smallest to largest.

5. 29, 25, 27 ________

6. 38, 58, 48 ________

7. ten more than 38 ________

8. ten less than 27 ________

9.
20 + 8 = ________
2 + 8 = ________

10.
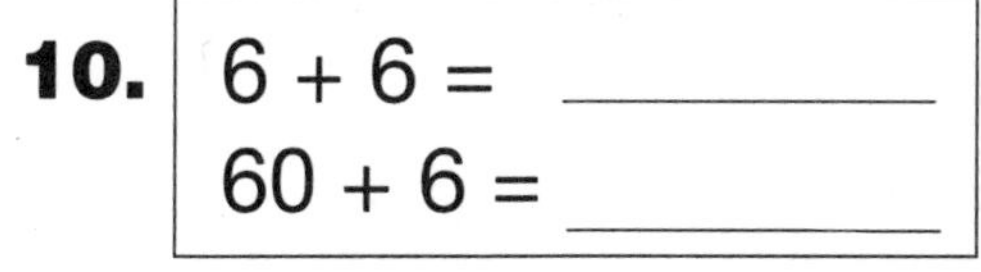
6 + 6 = ________
60 + 6 = ________

Go On What numbers are missing? 61, 51, 41, ______, ______, ______
What is the rule?

Date ____________ Name ____________________

What's in That Place? 5

STOP Don't solve yet! Circle two problems that may have the smallest answer.

Fill in the missing numbers.

1. 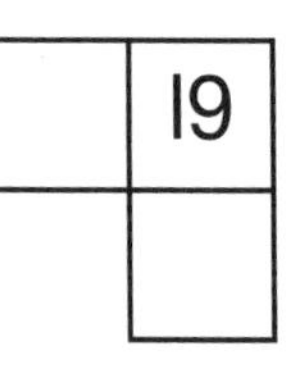

2. 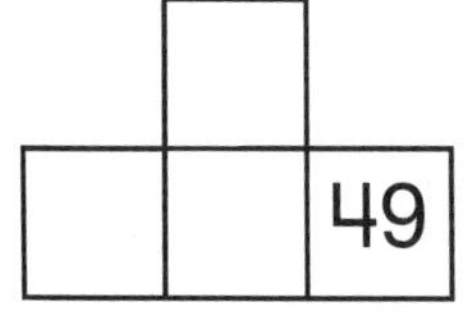

3. one more than 76 ________

4. one less than 63 ________

Order these numbers from smallest to largest.

5. 48, 44, 64 ________

6. 53, 13, 43 ________

7. ten more than 32 ________

8. ten less than 46 ________

9.
4 + 8 = ________
40 + 8 = ________

10.
60 + 7 = ________
6 + 7 = ________

Go On Write three numbers between 27 and 37.
How do you know they fit?

Date ____________ Name ____________________

What's in That Place? 6

STOP Don't solve yet! Circle a problem that may have the greatest answer.

Fill in the missing numbers.

1. 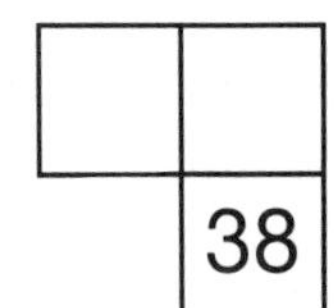

2. 36

3. one more than 85 ________

4. one less than 76 ________

Order these numbers from smallest to largest.

5. 57, 55, 54 ________

6. 29, 59, 39 ________

7. ten more than 43 ________

8. ten less than 27 ________

9.
30 + 9 = ________
3 + 9 = ________

10.
5 + 8 = ________
50 + 8 = ________

Go On Order these numbers from smallest to largest. 74, 68, 47, 82
How do you know?

Race to 40

Topic: Place Value of Tens and Ones

Object: Be first to display 40.

Groups: Pair players

Materials for each pair

- *Race to 40* gameboard, p. 114
- Mini Ten Frames sheet cut apart, p. 150
- 20 beans
- Number Cube (1–6), p. 145

Tip *Reverse the procedure by having pairs begin with four Ten Frames and "race to 0." The pairs remove beans according to the number rolled. If there are not enough beans, a Mini Ten Frame is exchanged for 10 beans that are placed in one of the larger Ten Frames. Some children will benefit by writing their new totals after each turn.*

Directions

1. The first pair rolls the Number Cube to determine how many beans to take. The pair places the beans in one of the Ten Frames in the lower portion of the gameboard. The pair states the amount shown.

2. The second pair follows the same procedure.

3. The first pair rolls again, adds that many beans to the Ten Frame, and verbalizes the total amount displayed.

4. As pairs continue to alternate turns and follow this procedure, they will fill one of their Ten Frames. Whenever this occurs, the pair removes the beans from the filled Ten Frames and appropriately places a Mini Ten Frame in the upper portion.

Example: A pair has 18 displayed and rolls 5. When the five beans are placed, a Ten Frame is filled and the other Ten Frame has three beans. The pair trades ten beans from the filled Ten Frame for a Mini Ten Frame which is placed over the "20" above. Then the pair states, "23 is our new total."

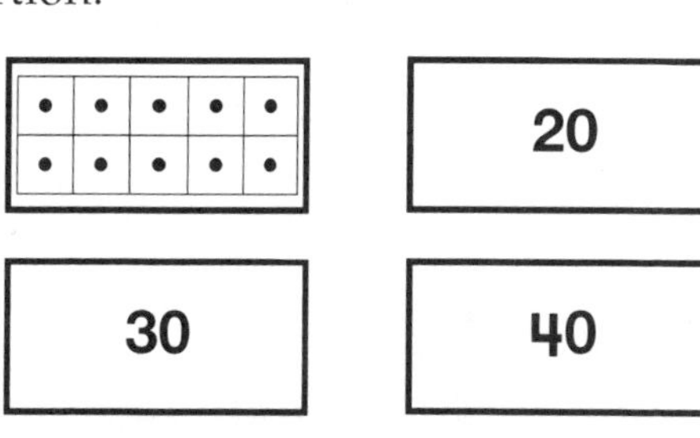

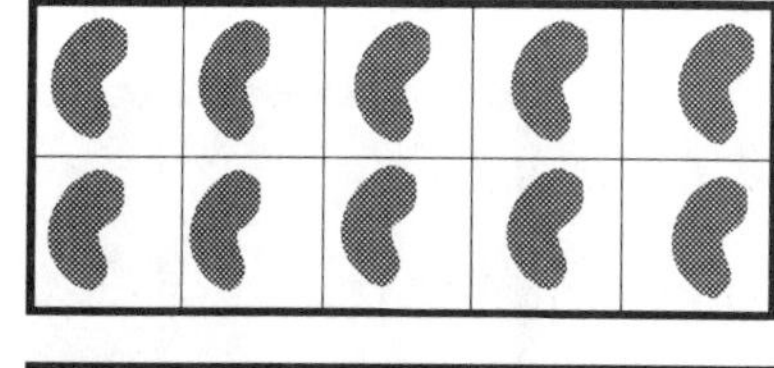

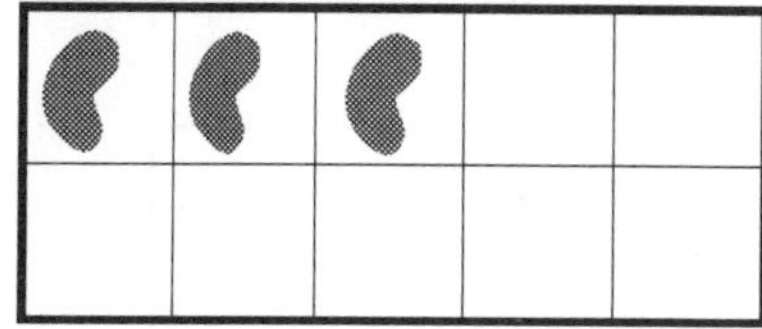

5. Play continues until one pair wins by displaying four Mini Ten Frames. Pairs are allowed to display more than 40.

Making Connections

Promote reflection and make mathematical connections by asking:

- What displayed totals allowed you to trade for a Mini Ten Frame on your next turn?

Race to 40

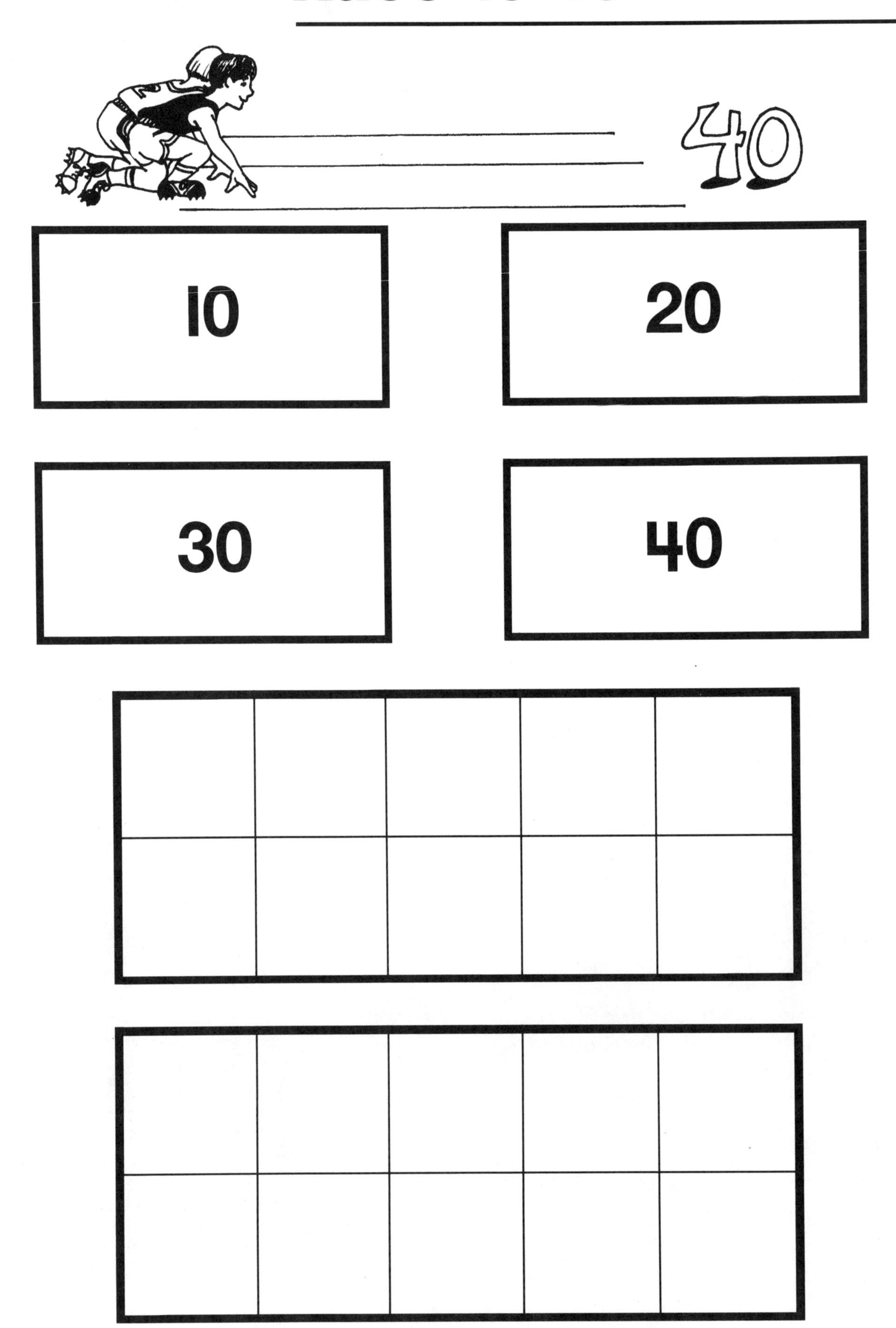

Claim All You Can

Topic: Number Relationships to 50

Object: Take the most counters off the gameboard.

Groups: Pair players

Materials for each group

- *Claim All You Can* gameboard, p. 116
- 20 transparent counters
- 2 game pieces (different kind for each pair)
- paper clip and pencil for spinner

Directions

1. Pairs place the twenty counters on the numbers labeled with a dot on the gameboard.

2. The first pair places a game piece on an uncovered number. Next the pair spins to determine where to move the game piece. The pair states aloud the action and the results. If the game piece lands on a number covered with a counter, the pair removes and keeps the claimed counter.

Example: A pair has covered 15 with a game piece. The pair spins "10 more." The pair moves the game piece to 25 and states, "25 is 10 more than 15." Since 25 is covered by a counter, the pair takes the counter that covered 25.

3. The second pair places the other game piece on an uncovered number and follows the same procedure. Be sure each pair verbalizes the action and the results of each turn.

4. Pairs alternate turns and continue to spin, move, verbalize, and claim counters. If the spinner indicates a move to a number not found on the gameboard, the pair spins again.

5. A pair wins a round by being the first pair to claim eleven counters.

1	2 •	3	4	5 •	6	7	8	9	10
11	12	13 •	14 •	15	16	17 •	18 •	19 •	20
21	22	23 •	24	25 •	26 •	27	28 •	29	30 •
31	32 •	33	34 •	35 •	36	37 •	38	39 •	40 •
41 •	42	43	44	45	46 •	47	48	49	50

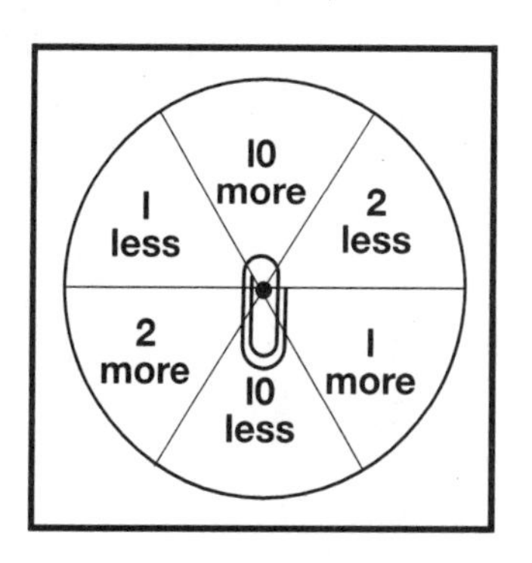

Making Connections

Promote reflection and make mathematical connections by asking:

- What are some of the good starting numbers?

Tips *Encourage number sense beyond 50 by having children play this game with the bottom half of a 100 Chart (51–100). A time-consuming but valuable option is to have pairs alternate turns and initially place the 20 counters.*

Claim All You Can

1	2 •	3	4	5 •	6	7	8	9	10
11	12	13 •	14 •	15	16	17 •	18 •	19 •	20
21	22	23 •	24	25 •	26 •	27	28 •	29	30 •
31	32 •	33	34 •	35 •	36	37 •	38	39 •	40 •
41 •	42	43	44	45	46 •	47	48	49	50

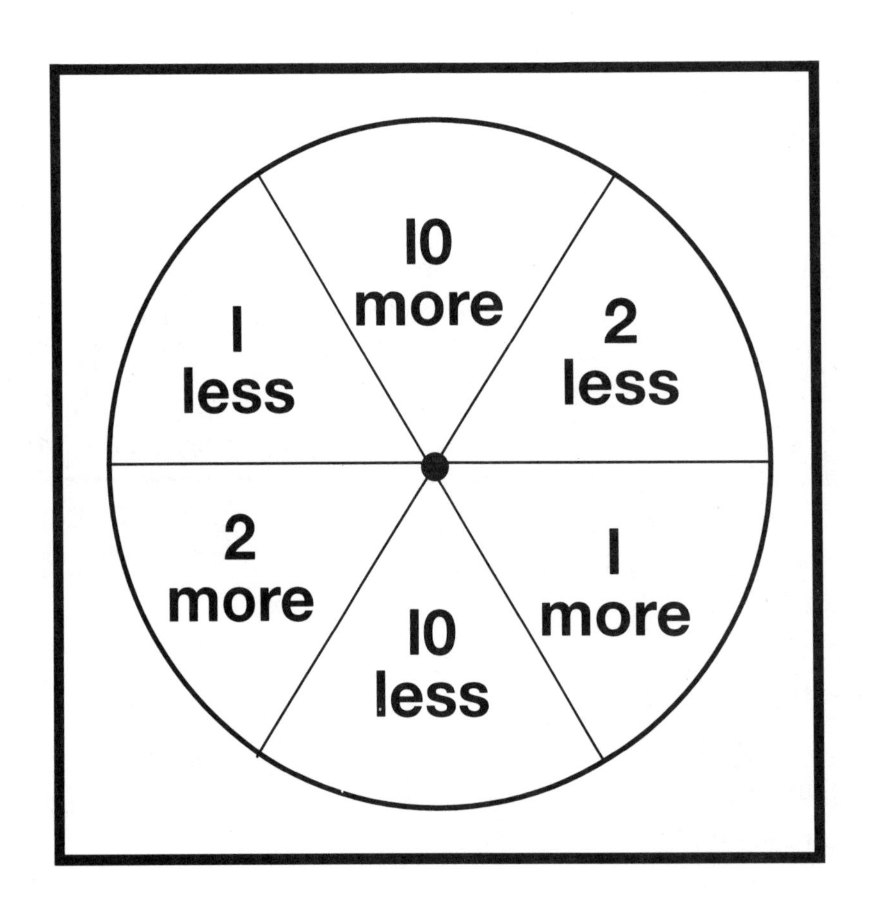

Low-High Spin

Topic: Place Value of Tens and Ones

Object: Order 2-digit numbers in an ascending sequence.

Groups: Pairs

Materials for each pair

- *Low-High Spin* gameboard, p. 118
- 100 Chart, p. 139
- 2 paper clips and a pencil for spinner
- inch squares of paper

Tip Make the experience more concrete by having pairs build the spun amounts with Mini Ten Frames and beans, p. 150.

Directions

1. In this game, a pair works cooperatively to order 2-digit numbers.
2. One member of the pair spins the tens spinner and the other member spins the ones spinner. The pair identifies the displayed 2-digit number and records it on one of the square pieces of paper.
3. Considering the choices offered by the spinners, the pair discusses and decides where to place their 2-digit number in a sequence of four possibilities. They place the numbered paper square in one of the gameboard boxes.
4. The pair continues spinning, recording, discussing, deciding, and placing the 2-digit numbers. If necessary, the pair is allowed to rearrange previously placed numbers.
5. Once a pair has ordered the four numbers, they use a 100 Chart to check the accuracy of their sequence. If the pair correctly completes an ascending sequence of four numbers, the pair wins that round.
6. Encourage children to play additional rounds.

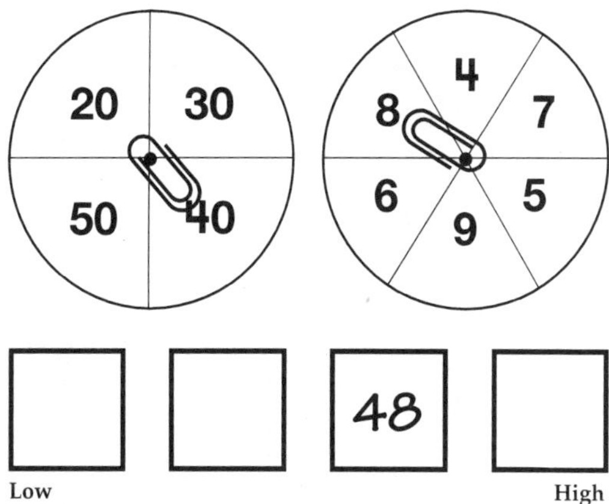

Making Connections

Promote reflection and make mathematical connections by asking:

- Which numbers fit well in the middle two boxes?

Low-High Spin

Date ____________ Name ____________

What Numbers Are Missing? I

Use the numbers shown on this 50 Chart to write the missing numbers in the empty spaces.

			4			7	8		
		13	14			17			20
	22	23					28	29	30
31	32	33	34			37	38		40
41			44	45			48		50

Fill in the missing numbers.

11		13		15			18		20
21	22				26	27		29	

31		33	34			37		39	40
41	42		44	45		47	48		50
51		53		55	56		58	59	60

Date ____________ Name ____________________

What Numbers Are Missing? II

Numbers are missing from these 100 Chart pieces. Write the missing numbers in the empty spaces.

41			44		46	47			50
51	52			55	56		58		60

61		63	64			67			70
	72	73			76	77	78		
81	82			85	86			89	
91			94			97			100

	72			75		77		79	
81			84						90
	92	93			96		98		

Date ____________________ Name ________________________________

Fill in the Pieces I

Each of these is a piece cut from a 50 Chart. Fill in the missing numbers.

27	

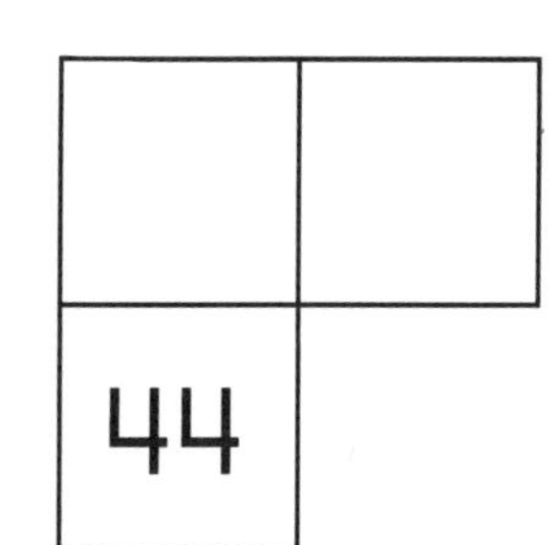

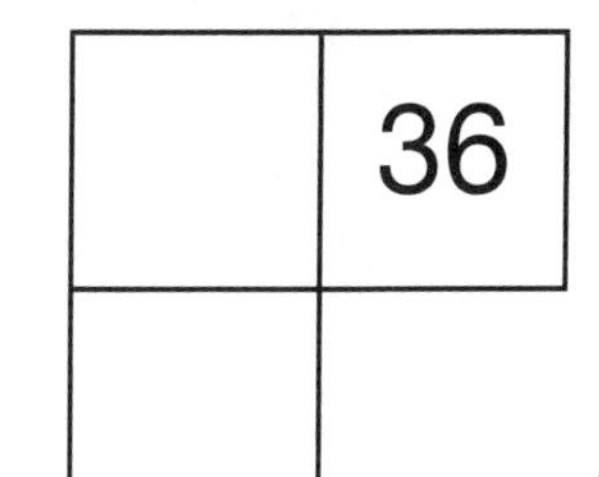

	32

	29

49	

	39

33	

	42

Date ____________________ Name ______________________________

Fill in the Pieces II

Each of these is a piece cut from a 100 Chart. Fill in the missing numbers.

23		
	44	

	35	
54		

46		
		58

73		

	66	

	83	

61		

	55	

		99

Date ______________ Name ______________________

Smallest and Largest I

Use 2 of the 3 digits to form the smallest and the largest two-digit number.

1. Use 1, 2, 4 ______ smallest ______ largest

2. Use 5, 2, 3 ______ smallest ______ largest

3. Use 2, 1, 5 ______ smallest ______ largest

4. Use 3, 2, 4 ______ smallest ______ largest

5. Use 1, 3, 5 ______ smallest ______ largest

6. Use 4, 2, 5 ______ smallest ______ largest

7. Use 3, 5, 4 ______ smallest ______ largest

8. Use 4, 1, 5 ______ smallest ______ largest

Date ____________ Name ____________________

Smallest and Largest II

Use 2 of the 3 digits to form the smallest and the largest two-digit number.

1. Use 4, 6, 7

These numbers are more / less than 10 apart.

________ smallest ________ largest

2. Use 6, 8, 9

These numbers are more / less than 10 apart.

________ smallest ________ largest

3. Use 3, 7, 5

These numbers are more / less than 50 apart.

________ smallest ________ largest

4. Use 2, 9, 8

These numbers are more / less than 50 apart.

________ smallest ________ largest

5. Use 8, 5, 6

These numbers are more / less than 10 apart.

________ smallest ________ largest

6. Use 6, 8, 3

These numbers are more / less than 30 apart.

________ smallest ________ largest

7. Use 4, 5, 9

These numbers are more / less than 30 apart.

________ smallest ________ largest

8. Use 4, 2, 8

These numbers are more / less than 30 apart.

________ smallest ________ largest

Money

Assumptions Money concepts have been taught for understanding, with an effort to enhance number sense. Concrete and visual models, such as coins and play money, have been used extensively. Children have practiced identifying coins and their values; skip counting by ones, fives, and tens; and finding the value of coin combinations. Only Sponges, Skill Checks, and Games are included in this money section. Because of appropriateness for first graders, Independent Activities have not been included. If your children need independent practice, refer to the money section in *Nimble with Numbers, Grades 2 and 3.*

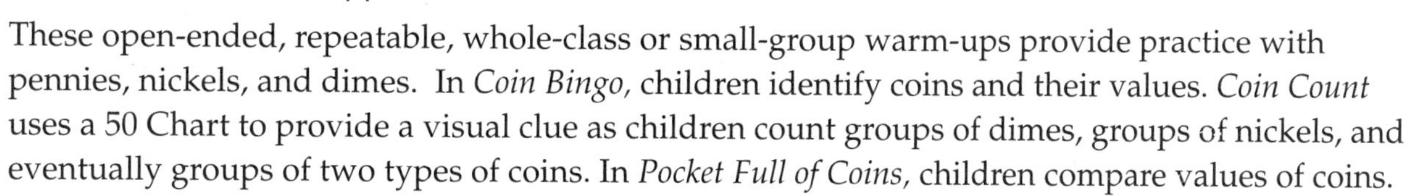

Section Overview and Suggestions

Sponges

Coin Bingo p. 126

Coin Count p. 127

Pocket Full of Coins pp. 128–129

These open-ended, repeatable, whole-class or small-group warm-ups provide practice with pennies, nickels, and dimes. In *Coin Bingo,* children identify coins and their values. *Coin Count* uses a 50 Chart to provide a visual clue as children count groups of dimes, groups of nickels, and eventually groups of two types of coins. In *Pocket Full of Coins,* children compare values of coins.

Skill Checks

Coins Count 1–4 pp. 130–131

These provide a way to help parents, children, and you to see children's improvement with identifying and combining coins. Copies of *Coins Count* may be cut in half so that each Check may be used at a different time. Remember to have children respond to STOP, the number sense task, before they solve the problems.

Games

Race to 20¢ pp. 132–133

Coin Draw pp. 134–136

These repeatable Games actively involve children in mentally combining coins. *Race to 20¢* requires children to make appropriate money exchanges. In *Coin Draw* children verbalize effective strategies as they discuss potential plays with their partners. Additional gameboards provide variations with different coins and combining three or four coins. Both Games are ones parents will enjoy playing repeatedly with their children.

Coin Bingo

Topic: Coin Recognition

Object: Identify a specific coin in order to get three-in-a-row.

Groups: Whole class or small group

Tips *Before using the Sponge, have children brainstorm all they know and observe about each of the coins. Children enjoy collaborating to create clues for future warm-ups.*

Materials

- prepared clues on cards
- *Three-in-a-Row* form for each child, p. 17
- 9 coins for each child (3 pennies, 3 nickels, 3 dimes)
- 9 counters for each child

Directions

1. Children randomly arrange their nine coins on a *Three-in-a-Row* form to create a unique playing board.
2. The leader draws and reads a clue card.

   Example: "Find a coin that is worth ten pennies."
3. Children individually identify the described coin and cover one of the described coins with a counter. Before continuing on to the second clue, children agree on the correct response to the first clue.
4. The leader continues to follow this procedure of giving clues and having children locate and cover the corresponding coins.

   Clues "Find a coin that . . ."

   1¢ is copper in color
   5¢ is thickest of these three coins
   10¢ has ridges on the edge
   1¢ is worth the least amount, but is not smallest in size
   5¢ is worth more than a penny, less than a dime
   10¢ is worth two nickels
   1¢ has Abraham Lincoln's picture on it
   5¢ is worth five pennies
   10¢ is thinnest of these three coins

5. When children have three covered coins in a row, they call out, "Three-in-a-row." The Game continues until all children have at least one "three-in-a-row." (Some children may have several three-in-a-rows.)
6. Children seem to enjoy repeated rounds of this Sponge.

Making Connections

Promote reflection and make mathematical connections by asking:

- What did you do to figure out the clue?

Coin Count

Topic: Adding Like Coins

Object: Skip count on 50 Chart to represent coin values.

Groups: Whole class or small group

Materials

- transparency of 50 Chart, p. 138
- transparent coins, p. 140

Directions

1. The leader announces, "Let's count dimes. What's the value of one dime? Two dimes? Count aloud as each coin is placed on the 50 Chart."
2. The leader places four dimes, one at a time, on the 50 Chart, one each on top of the numbers 10, 20, 30, 40. As each dime is placed, children count, "10, 20, 30, 40." The leader asks, "When dimes are placed, what counting pattern did you use?" (counting by tens)
3. The leader repeats the procedure with different amounts of dimes.
4. The leader begins again with nickels, and announces "Let's count nickels. What's the value of one nickel? Two nickels? Count aloud as each nickel is placed on the 50 Chart."
5. The leader places seven nickels, one at a time, on the 50 Chart, one each on top of multiples of 5 (from 5 through 35). As each nickel is placed, children count, "5, 10, 15, 20, 25, 30, 35." The leader asks, "When nickels are placed, what counting pattern did you use?" (counting by fives)
6. The leader repeats the procedure with different amounts of nickels.
7. After counting with only one type of coin, children count totals using two different types of coins. As the number and types of coins are varied, the steps are repeated. Count the coins with the larger values first (dimes, then nickels or pennies).

   Example: Place 3 dimes (one each on 10, 20, 30), then 4 pennies (one each on 31, 32, 33, 34) on the 50 Chart, counting aloud "10, 20, 30, 31, 32, 33, 34."

Tips *When using two different types of coins, children usually experience more success if dimes and pennies are counted together first, then nickels and pennies, then dimes and nickels. If amounts greater than 50¢ are counted, use a 100 Chart in place of the 50 Chart.*

1	2	3	4	5	6	7	8	9	
11	12	13	14	15	16	17	18	19	
21	22	23	24	25	26	27	28	29	
31	32	33	34	35	36	37	38	39	
41	42	43	44	45	46	47	48	49	50

Making Connections

Promote reflection and make mathematical connections by asking:

- Did the 50 Chart help you count coins more easily? Please explain.

Pocket Full of Coins

Topic: Adding Like Coins

Object: Total coins and predict value for a related representation.

Groups: Whole class or small group

Materials

- transparency of *Pocket Full of Coins* form cut apart, p. 129
- transparent coins, p. 140

Directions

1. The leader displays the smallest pocket and places one penny inside as a reference. The leader asks, "How many pennies will it take to fill the pocket?" Children discuss and estimate.
2. The leader selects a child to help fill the pocket with pennies to check out classmate's estimates. The leader asks, "What is the value of these pennies?" Children discuss. Then the leader points to each penny as children count the value by ones.
3. The leader records the value and removes all pennies from the pocket.
4. Next the leader places one dime inside the pocket as a reference. The leader asks, "How many dimes will it take to fill the pocket? What will be the value of those dimes?" The leader should remind children to think about the value of the pennies that they just counted. Children discuss and estimate.
5. The leader selects a child to help fill the pocket with dimes to check out classmate's estimates. The leader asks, "What is the value of these dimes?" Children discuss. Then the leader points to each dime as children count the value by tens.
6. The leader records the value and removes all dimes from the pocket.
7. The leader places one nickel inside the pocket and repeats the procedure.
8. For future rounds, repeat the procedure with different-sized pockets.

Tips *Some children will appreciate having their own copies of the pockets. Duplicate the* Pocket Full of Coins *form for children to use at tables with coins. Children will enjoy drawing their own shapes to fill with coins and compare values.*

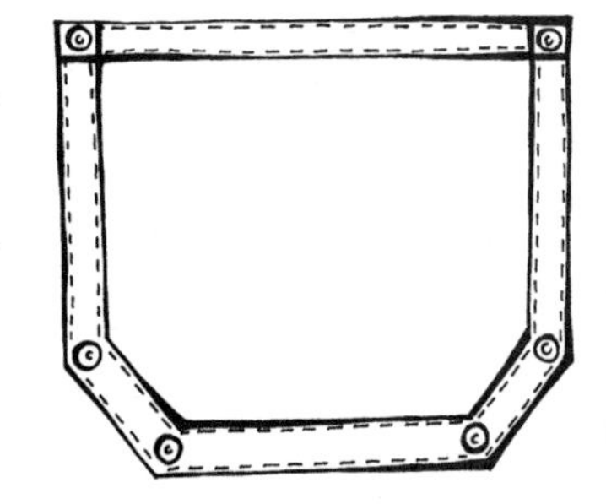

Making Connections

Promote reflection and make mathematical connections by asking:

- How did knowing the value of pennies help you figure the value of dimes? Nickels?

Pocket Full of Coins

Date ______________ Name ________________________

Coins Count 1

Don't start yet! Circle a problem that may have the greatest answer.

1. 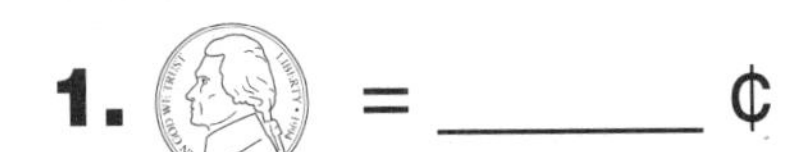 = ______ ¢

2. = ______ ¢

3.  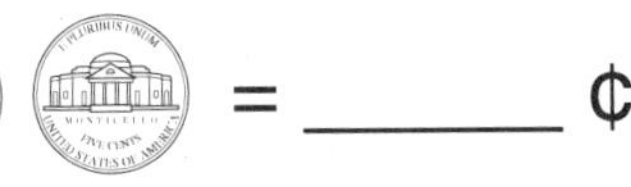 = ______ ¢

4. Use 2 coins to make 6¢.

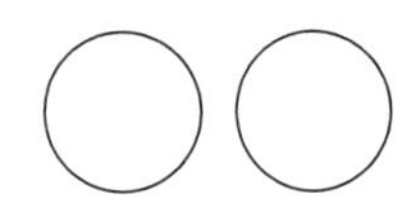

5. Show 10¢ two ways.

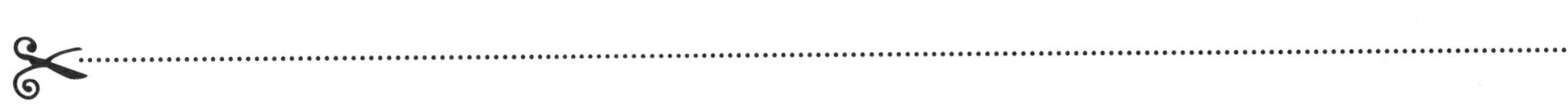

Date ______________ Name ________________________

Coins Count 2

Don't start yet! Circle a problem that may have an even answer.

1. = ______ ¢

2.  = ______ ¢

3.   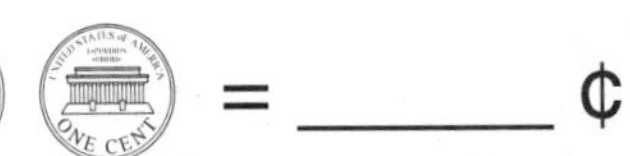 = ______ ¢

4. Use 2 coins to make 15¢.

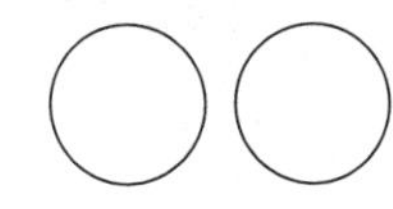

5. Show 10¢ two ways.

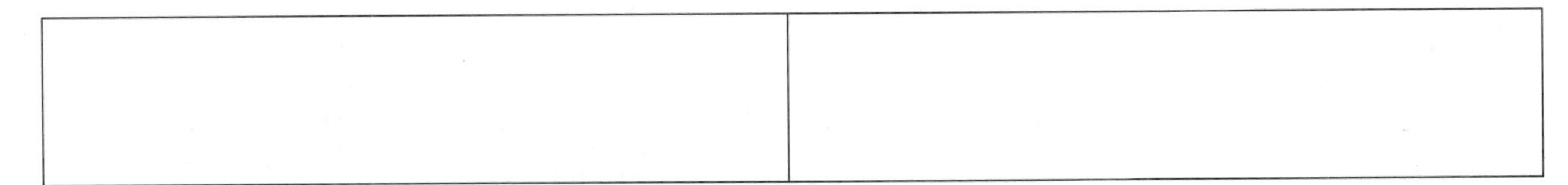

Date ____________ Name ____________________

Coins Count 3

STOP Don't start yet! Circle a problem that may have the smallest answer.

1. 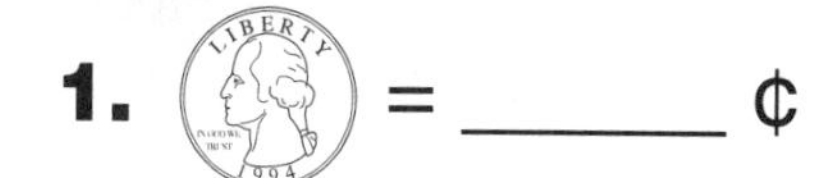 = ______ ¢

2.  = ______ ¢

3.  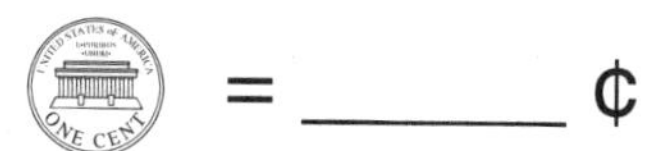 = ______ ¢

4. Use 2 coins to make 10¢.

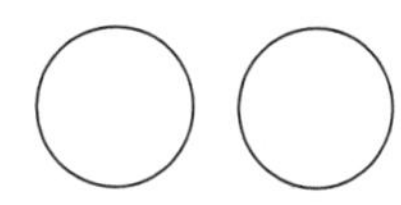

5. Show 25¢ two ways.

Date ____________ Name ____________________

Coins Count 4

STOP Don't start yet! Circle a problem that may have an odd answer.

1. = ______ ¢

2.  = ______ ¢

3.   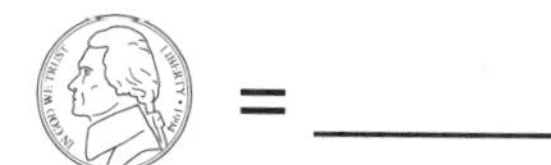 = ______ ¢

4. Use 2 coins to make 11¢.

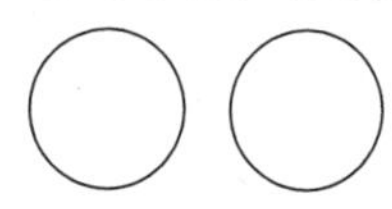

5. Show 20¢ two ways.

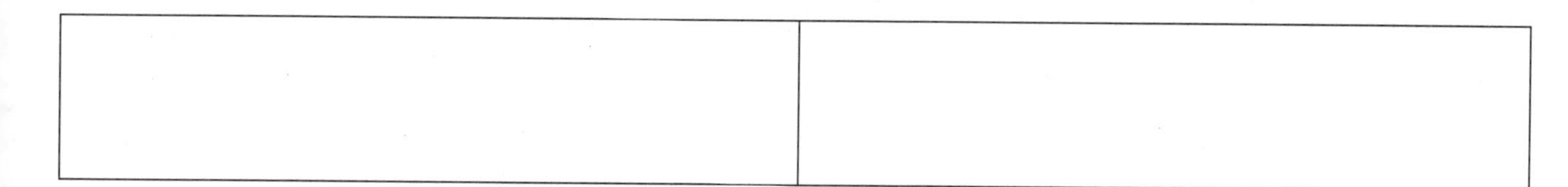

Race to 20¢

Topic: Exchanging and Adding Pennies, Nickels, and Dimes

Object: Reach 20¢.

Groups: 2 players

Tip *Some children might find it helpful to record their turns and keep track of their totals.*

Materials for each group

- *Race to 20¢* gameboard for each player, p. 133
- special Number Cube (1-1-2-2-3-3), p. 147
- 10 counters
- real or play Coins (pennies, nickels, and dimes), p. 140

Directions

1. The first player rolls the Number Cube. The number rolled indicates the number of pennies awarded for that turn. The player covers the rolled amount on his or her gameboard and states the accumulated value. After accumulating 5 pennies, a player must exchange them for a nickel.
2. The second player rolls the Number Cube, indicates the value of the roll on his or her gameboard, and states the accumulated value.
3. Players continue to alternate turns and follow the same procedure. Players must exchange coins when appropriate (five pennies for a nickel, two nickels for a dime). When a player accumulates one dime, the dime is placed in the treasure chest and that player is halfway to a winning round. Players win when they have accumulated two dimes. If players have the same number of turns, it is possible both players could win.
4. Since exchanging coins is worthwhile practice, players are encouraged to play additional rounds.

Making Connections

Promote reflection and make mathematical connections by asking:

- How did you know when to trade your coins for a different coin?
- If you were to redesign the number cube, how would you change it? Why?

Race to 20¢

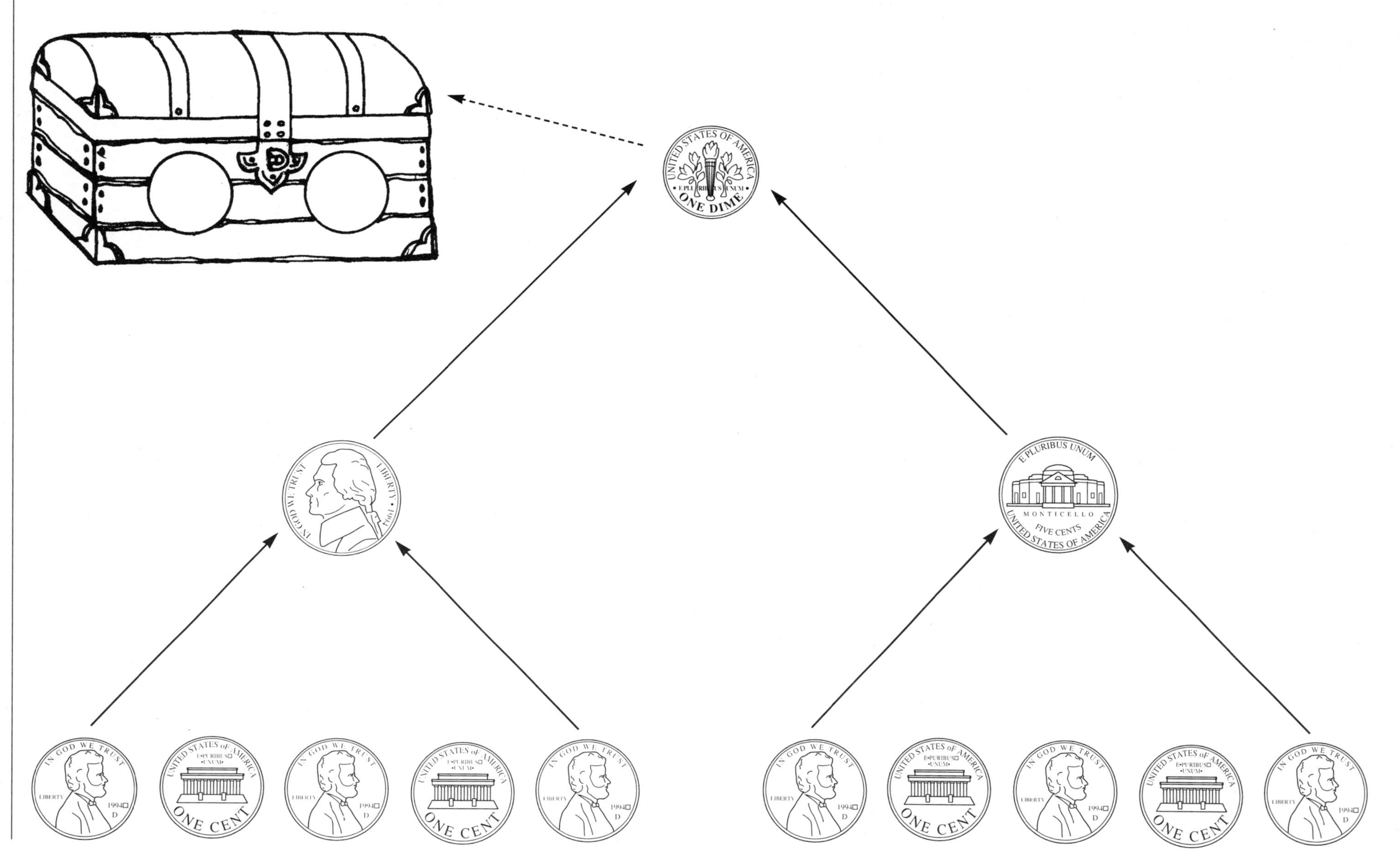

Coin Draw

Topic: Adding Coins

Object: Combine coins to make three-in-a-row.

Groups: Pair players

Materials for each group

- *Coin Draw A* gameboard, p. 135
- counters (different type for each pair)
- 3 dimes and 3 pennies in a sock (or hidden container)

Directions

1. Children empty the sock and verify there are three dimes and three pennies, and return the six coins to the sock. One member of the first pair reaches inside the sock and without looking removes three coins. The pair identifies the value of the three coins.
2. The first pair discusses and decides which one corresponding total on the gameboard to cover with their counter, covers the amount, and returns the drawn coins to the sock.
3. One member of the second pair draws three coins from the sock, determines the value of the three coins, and selects and covers one number with their type of counter.
4. Pairs continue to alternate turns, removing three coins, totaling the value of coins drawn, and selecting and covering one number with their counter.
5. If a drawn-coin combination does not result in any available total to cover, a pair returns the coins and draws once more.
6. The first pair to have three counters in a row, horizontally, vertically, or diagonally, wins.
7. The *Coin Draw B* gameboard, p. 135, combines nickels with pennies. The sock contains three nickels and three pennies, and pairs draw three coins and follow the same procedure to make the totals.

Tip *To play with the Four-Coin Draw gameboards, p. 136, children remove four coins from the sock on each turn. The sock for gameboard A should contain four dimes and four pennies; the sock for gameboard B should contain four nickels and four pennies.*

12¢	21¢	3¢
3¢	30¢	12¢
12¢	21¢	30¢

Making Connections

Promote reflection and make mathematical connections by asking:

- What helped you get your counters in a row?
- How did you select which number to cover with your counter?

Coin Draw A

12¢	21¢	3¢
3¢	30¢	12¢
12¢	21¢	30¢

Coin Draw B

11¢	15¢	11¢
15¢	3¢	7¢
3¢	7¢	11¢

Four Coin Draw A

13¢	4¢	40¢
31¢	22¢	13¢
22¢	40¢	4¢

Four Coin Draw B

12¢	20¢	16¢
16¢	12¢	4¢
4¢	8¢	20¢

Blackline Masters

50 Chart

100 Chart

Coins

Counting Cards

Digit Cards

Digit Squares

Dot Cubes (1–6, 1-2-3-4-5-Choose)

Dot Pattern Cards

Dot Pattern Squares

Number Cubes (1–6, blank)

Spinners (1–6, blank)

Ten Frames

Mini Ten Frames

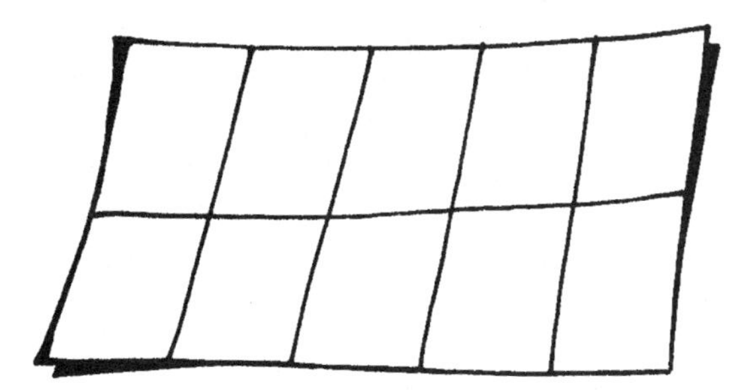

50 Chart

1	2	3	4	5	6	7	8	9	10
11	12	13	14	15	16	17	18	19	20
21	22	23	24	25	26	27	28	29	30
31	32	33	34	35	36	37	38	39	40
41	42	43	44	45	46	47	48	49	50

1	2	3	4	5	6	7	8	9	10
11	12	13	14	15	16	17	18	19	20
21	22	23	24	25	26	27	28	29	30
31	32	33	34	35	36	37	38	39	40
41	42	43	44	45	46	47	48	49	50

100 Chart

1	2	3	4	5	6	7	8	9	10
11	12	13	14	15	16	17	18	19	20
21	22	23	24	25	26	27	28	29	30
31	32	33	34	35	36	37	38	39	40
41	42	43	44	45	46	47	48	49	50
51	52	53	54	55	56	57	58	59	60
61	62	63	64	65	66	67	68	69	70
71	72	73	74	75	76	77	78	79	80
81	82	83	84	85	86	87	88	89	90
91	92	93	94	95	96	97	98	99	100

Coins

IN GOD WE TRUST
LIBERTY
1994
D
UNITED STATES OF AMERICA
E PLURIBUS UNUM
ONE CENT
MONTICELLO
FIVE CENTS
ONE DIME
QUARTER DOLLAR

Counting Cards

1	2	3	4
5	6	7	8
9	10	11	12
13	14	15	16
17	18	19	20

Digit Cards

0		
1	2	3
4	5	6
7	8	9

Digit Squares

0	1	2	3	4
5	6	7	8	9

0	1	2	3	4
5	6	7	8	9

0	1	2	3	4
5	6	7	8	9

Dot Cubes

Cut solid lines. Fold on dotted lines.

Choose

Dot Pattern Cards

Dot Pattern Squares

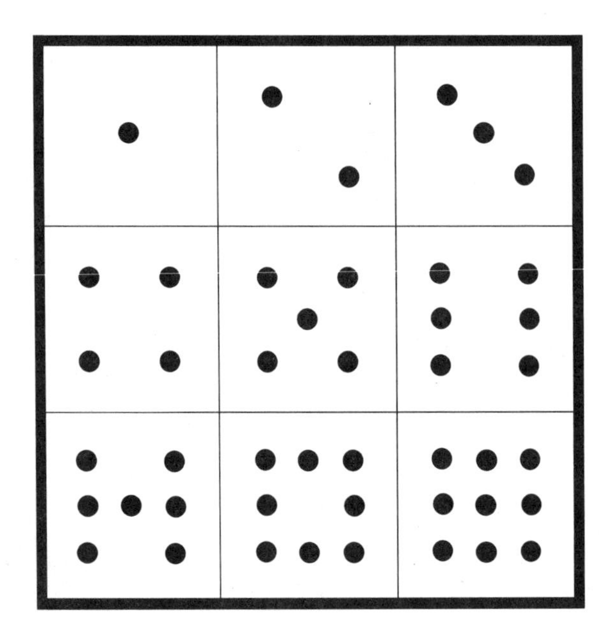

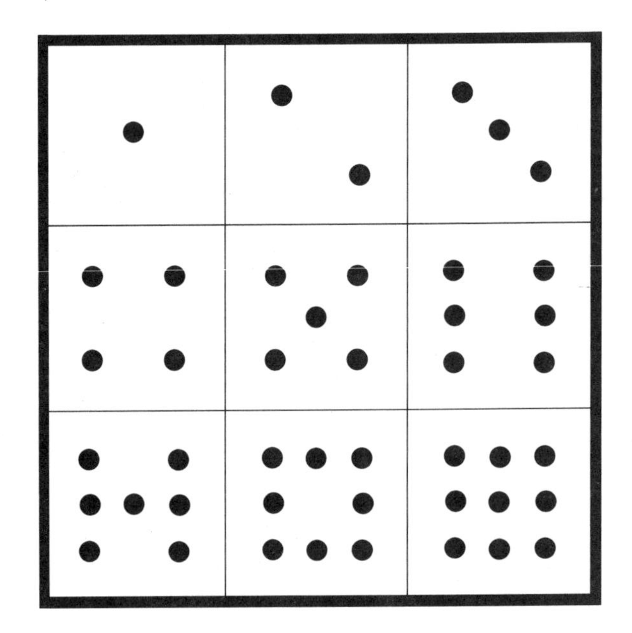

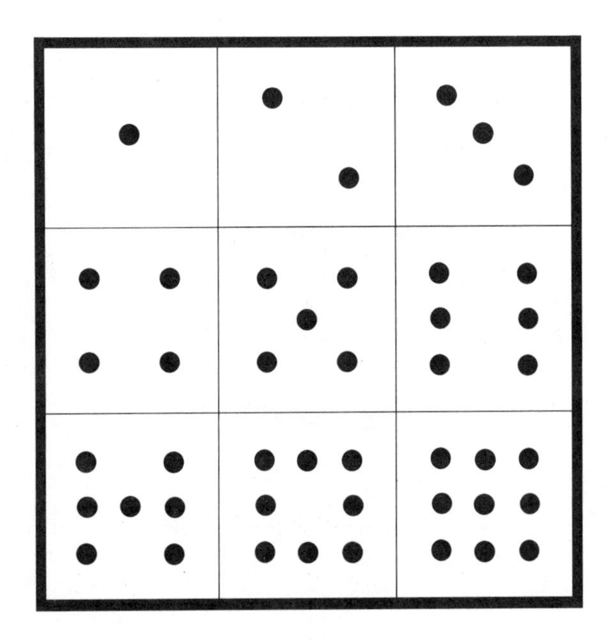

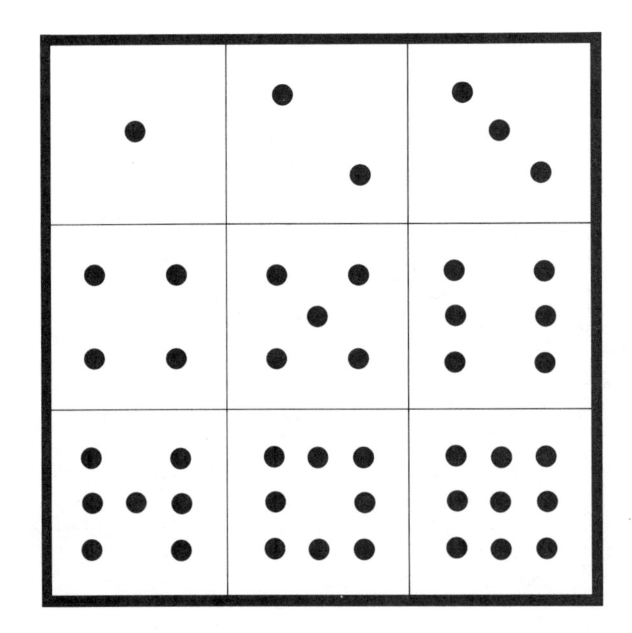

Number Cubes

Cut solid lines. Fold on dotted lines.

3

1 5

2 4

6

Spinners

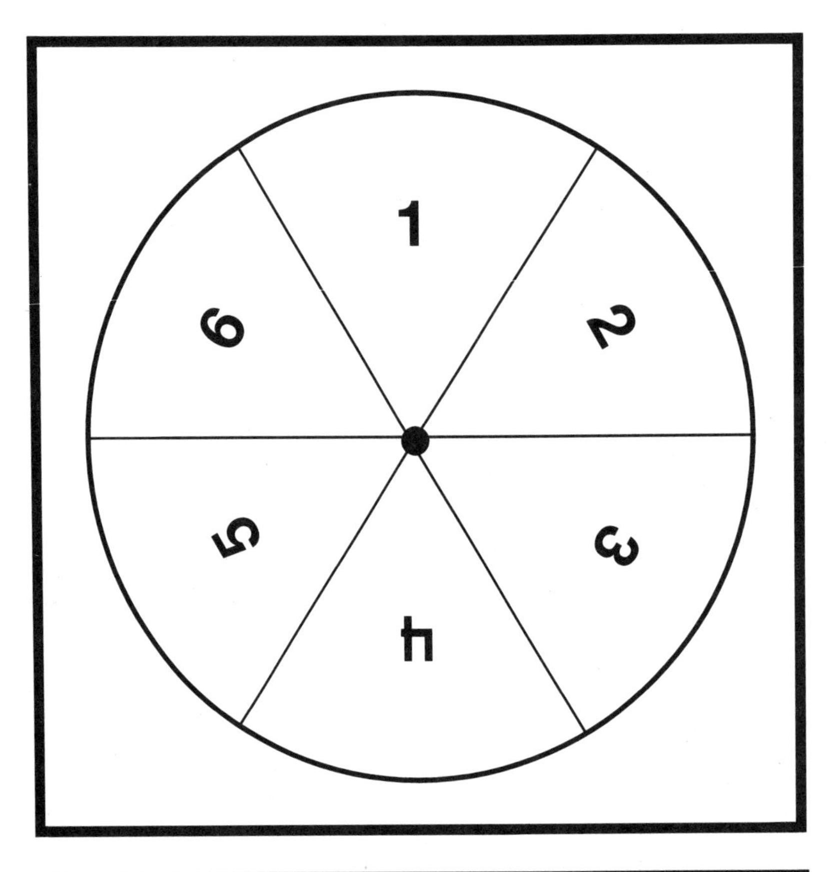

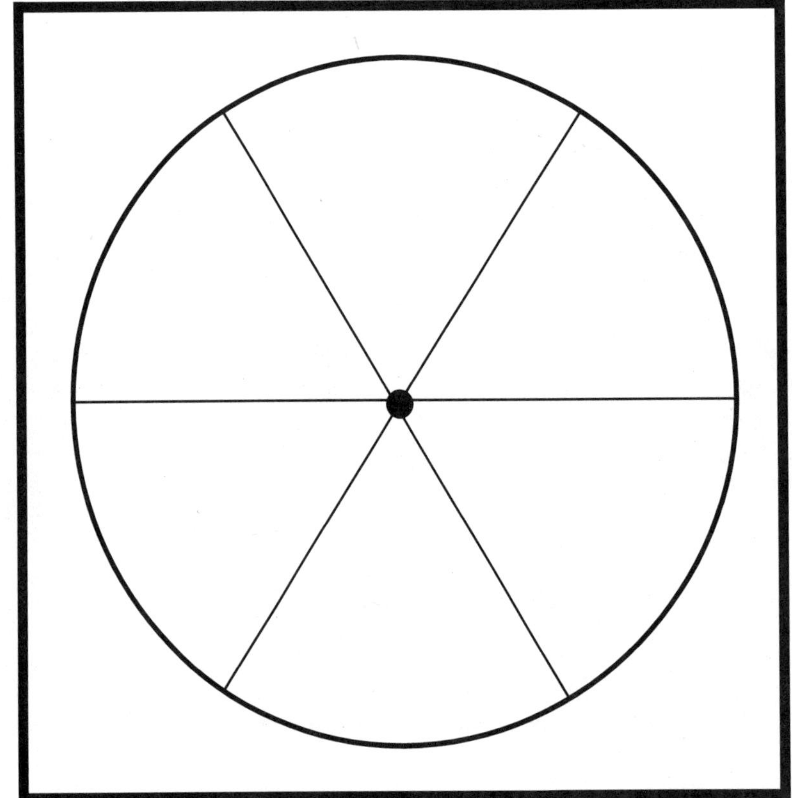

Ten Frames

Mini Ten Frames

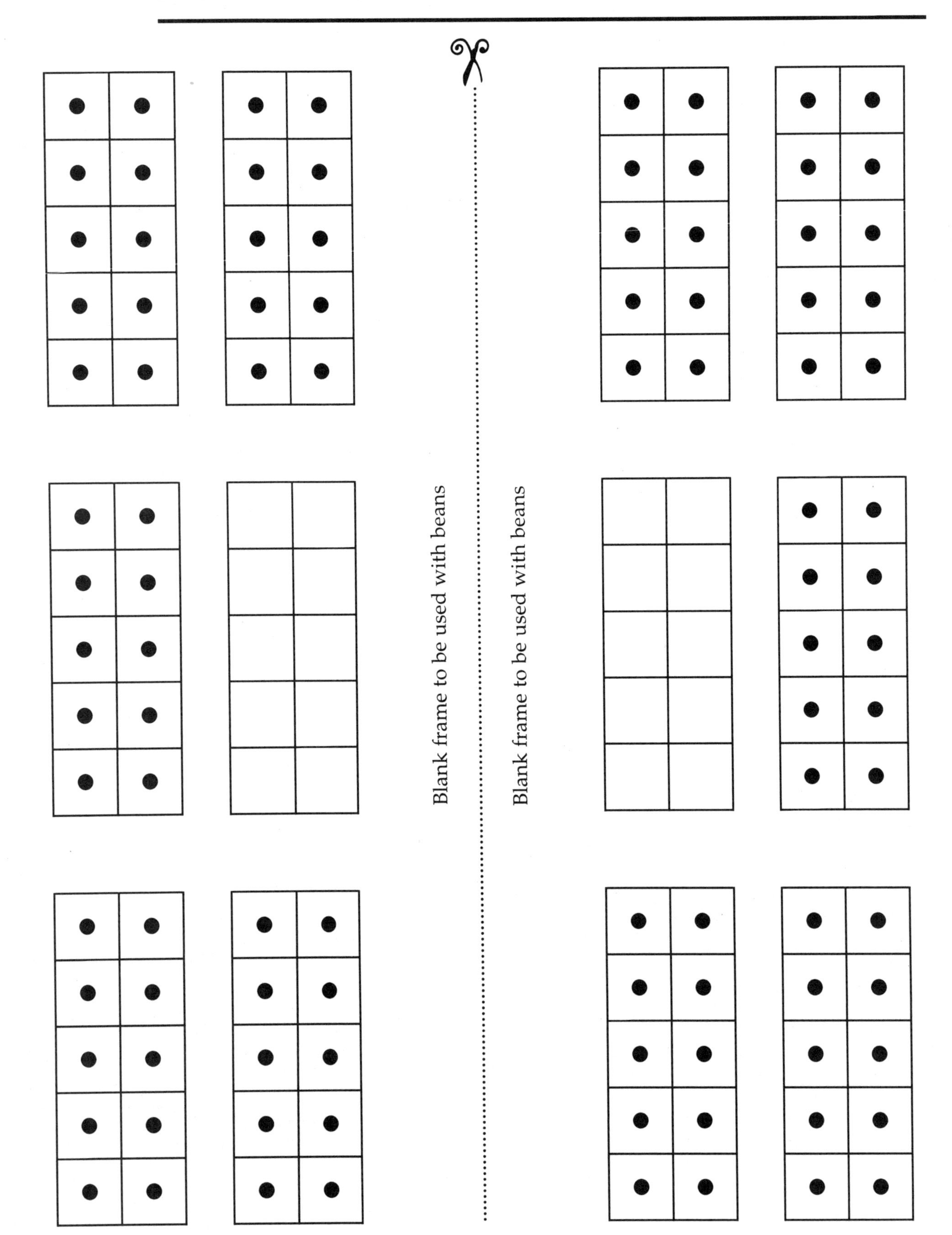

Nimble with Numbers Answer Key

p.18 ***Count On 1***

1) 6, 7 2) 5 3) 18 4) 15 5) 6 6) 8 7) 15 8) 12 9) 2,4 10) 11, 13
Go On 15, 13, 11, 9

Count On 2

1) 2, 4 2) 6 3) 12 4) 13 5) 7 6) 8 7) 11 8) 17 9) 6, 8 10) 17, 19
Go On 7, 9; 2, 4

p. 19 ***Count On 3***

1) 4,5 2) 8 3) 16 4) 11 5) 9 6) 6 7) 13 8) 15 9) 4, 6 10) 13, 15
Go On 18, 16, 14, 12

Count On 4

1) 8, 9 2) 4 3) 19 4) 18 5) 8 6) 10 7) 16 8) 18 9) 5, 7 10) 14, 16
Go On Answers will vary.

p. 20 ***Count On 5***

1) 8, 9 2) 9 3) 13 4) 14 5) 7 6) 8 7) 19 8) 10 9) 3, 5 10) 12, 14
Go On 17, 16, 14, 13

Count On 6

1) 5, 7 2) 7 3) 18 4) 11 5) 5 6) 10 7) 16 8) 17 9) 7, 9 10) 16, 18
Go On Answers will vary.

p. 27 ***More or Less I***

Answers will vary.

p. 28 ***More or Less II***

Answers will vary.

p. 29 ***Before, Between, After I***

Answers will vary.

p. 30 ***Before, Between, After II***

Answers will vary.

p. 31 ***Which Numbers Fit? I***

1) 7-10 2) 0-3 3) 4-10 4) 0-7 5) 8-10 6) 0-8 7) 6-10 8) 0-9 9) 1-10 10) 0-1

p. 32 ***Which Numbers Fit? II***

1) 14-20 2) 10-14 3) 17-20 4) 10-16 5) 12-20 6) 10-19 7) 15-20 8) 10-11 9) 14-20 10) 10-17

p. 39 ***Just the Facts 1***

1) 5 2) 8 3) 10 4) 11 5) 6 6) 9 7) 9 8) 12 9) 11 10) 14
Go On 8, 10, 12; The numbers increase by 2.

Just the Facts 2

1) 6 2) 9 3) 9 4) 12 5) 8 6) 10 7) 10 8) 11 9) 12 10) 13
Go On Answers will vary.

p. 40 ***Just the Facts 3***

1) 7 2) 9 3) 12 4) 11 5) 7 6) 10 7) 9 8)12 9) 11 10) 13
Go On 7, 9, 11; The numbers increase by 2.

Just the Facts 4

1) 8 2) 10 3) 12 4) 12 5) 9 6) 9 7) 10 8) 11 9) 12 10) 14
Go On Answers will vary.

p. 41 ***Just the Facts 5***

1) 8 2) 9 3) 11 4) 11 5) 7 6) 10 7) 9 8)12 9) 11 10) 13
Go On 10, 13, 16; The numbers increase by 3.

Just the Facts 6

1) 9 2) 10 3) 12 4) 12 5) 6 6) 10 7) 10 8) 11 9) 12 10) 14
Go On Answers will vary.

p. 53 *Seeking Sums Practice I*

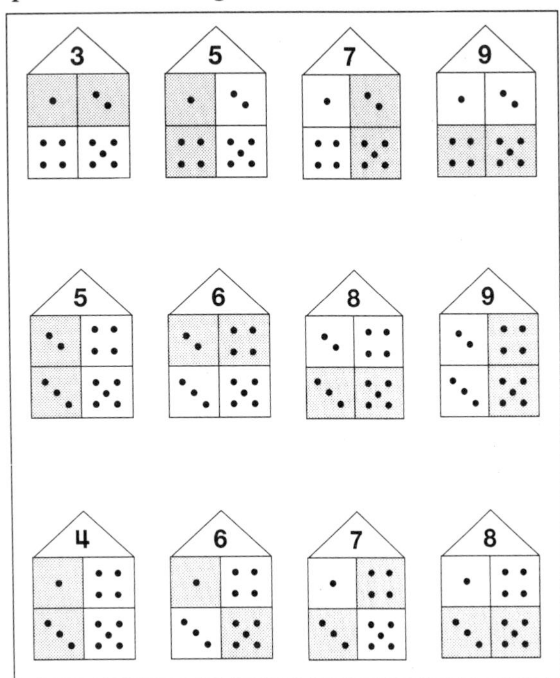

p. 54 *Seeking Sums Practice II*

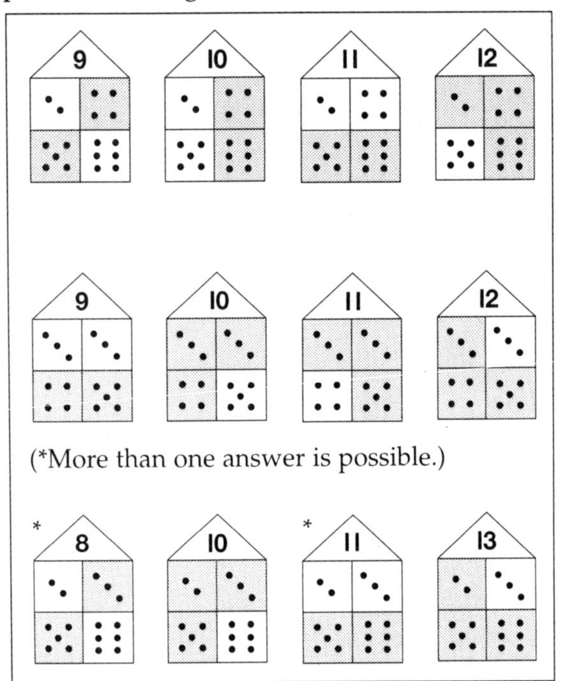

p. 55 *Roll and Fill I*

Answers will vary.

p. 56 *Roll and Fill II*

Answers will vary.

p. 57 *Neighbor Sums I*

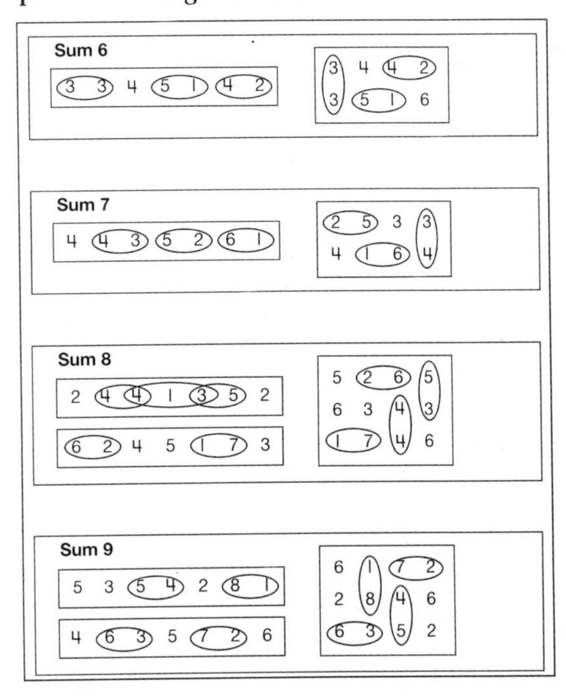

p. 58 *Neighbor Sums II*

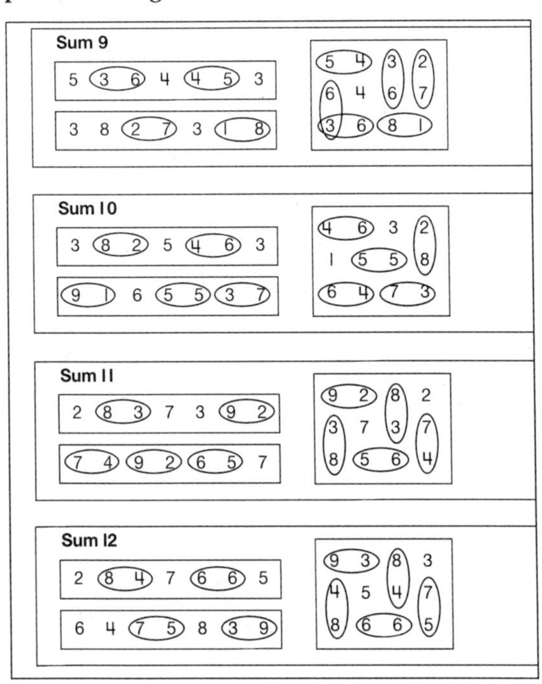

p. 64 *Just the Facts 7*

1) 3 2) 2 3) 6 4) 6 5) 2 6) 6 7) 7 8) 4 9) 2 10) 1

Go On 5; The numbers decrease by 2.

Just the Facts 8

1) 1 2) 3 3) 7 4) 5 5) 5 6) 6 7) 7 8) 7 9) 1 10) 1

Go On Answers will vary.

p. 65 ***Just the Facts 9***

1) 2 2) 3 3) 4 4) 1 5) 3 6) 4 7) 4 8) 2 9) 3 10) 1

Go On 6, 4; The numbers decrease by 2.

Just the Facts 10

1) 1 2) 5 3) 8 4) 1 5) 7 6) 7 7) 5 8) 1 9) 4 10) 1

Go On Answers will vary.

p. 66 ***Just the Facts 11***

1) 3 2) 4 3) 5 4) 0 5) 1 6) 3 7) 8 8) 3 9) 5 10) 1

Go On 4; The numbers decrease by 3.

Just the Facts 12

1) 2 2) 2 3) 9 4) 4 5) 4 6) 4 7) 8 8) 5 9) 6 10) 1

Go On Answers will vary.

p. 75 ***Before, Between, After I***

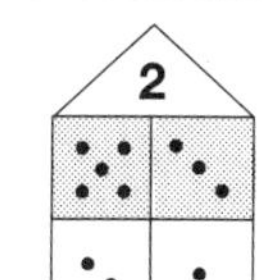
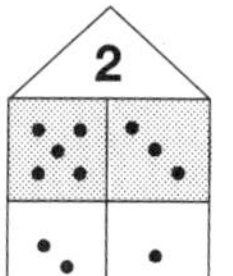
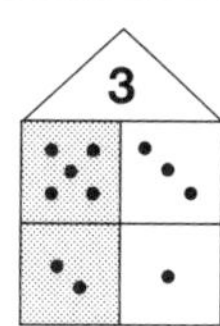
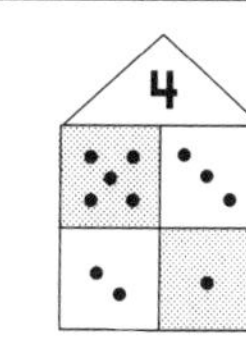

(*More than one answer is possible.)

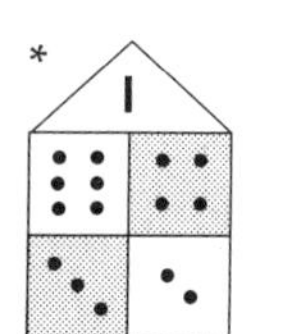
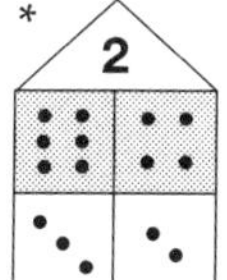
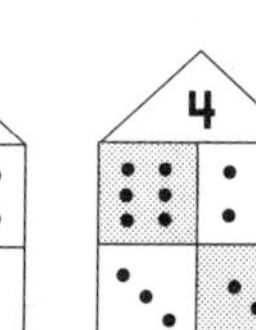
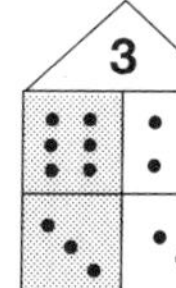
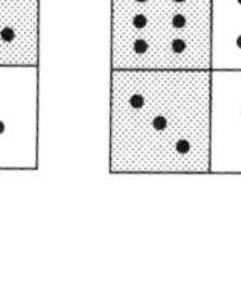

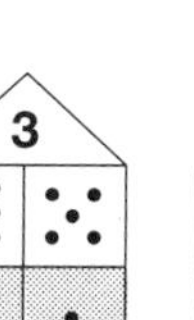
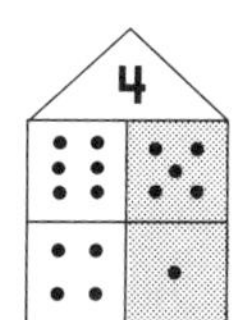
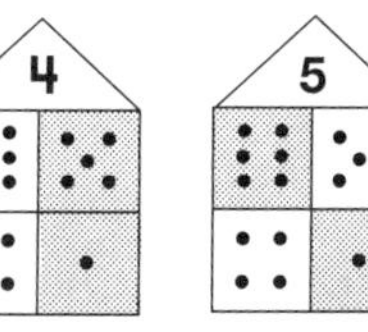

p. 76 ***Before, Between, After II***

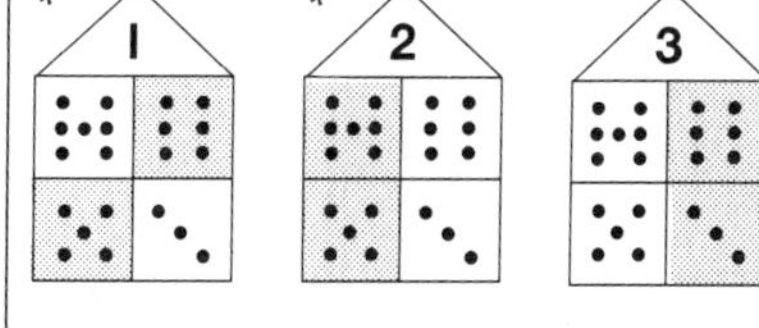

(*More than one answer is possible.)

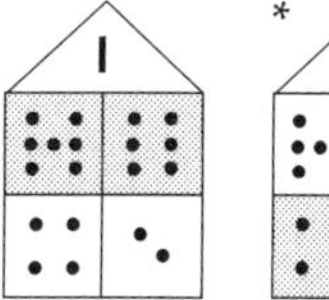
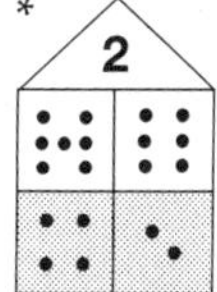
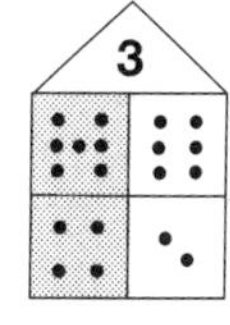
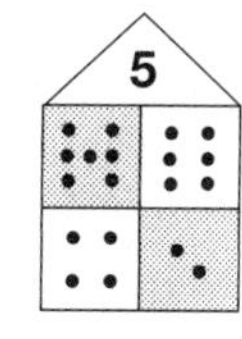

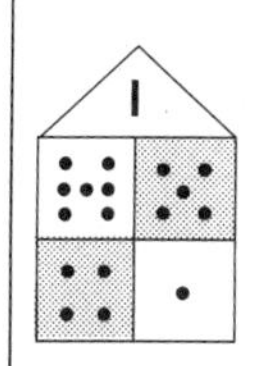
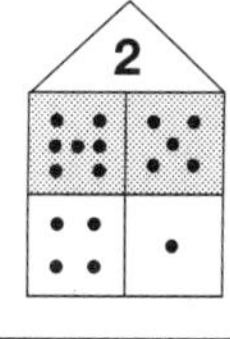
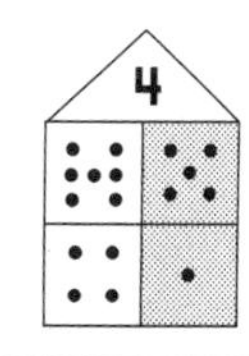
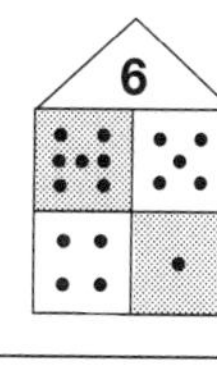

. 77 ***Roll and Fill Differences I***

Answers will vary.

p. 78 ***Roll and Fill Differences II***

Answers will vary.

. 79 ***Equation Hunt I*** Sample answers are given.

1) 4 – 3 = 1 2) 2 – 1 = 1 3) 4 – 4 = 0 4) 4 – 2 = 2 5) 1 – 1 = 0

6) 3 – 2 = 1 7) 5 – 3 = 2 8) 8 – 8 = 0 9) 6 – 4 = 2 10) 6 – 6 = 0

11) 6 – 3 = 3 12) 7 – 2 = 5

. 80 ***Equation Hunt II*** Sample answers are given.

1) 6 – 5 = 1 2) 8 – 4 = 4 3) 5 – 4 = 1 4) 7 – 4 = 3 5) 7 – 6 = 1

6) 8 – 3 = 5 7) 8 – 7 = 1 8) 9 – 6 = 3 9) 8 – 6 = 2 10) 9 – 7 = 2

11) 9 – 8 = 1 12) 9 – 4 = 5

. 87 ***Mixed Facts 1***

1) 3 2) 9 3) 7 4) 6 5) 12 6) 8 7) – 8) + 9) 11 10) 2

Go On Answers will vary.

Mixed Facts 2

1) 3 2) 9 3) 9 4) 2 5) 13 6) 4 7) + 8) – 9) 8 10) 1

Go On 12 – 3 = 9; 12 + 3 = 15

p. 88	***Mixed Facts 3***									
	1) 5	2) 11	3) 9	4) 6	5) 13	6) 7	7) +	8) –	9) 8	10) 1
	Go On 8, 14, 17; The numbers increase by 3.									
	Mixed Facts 4									
	1) 5	2) 11	3) 6	4) 3	5) 14	6) 5	7) –	8) +	9) 7	10) 2
	Go On Answers will vary.									
p. 89	***Mixed Facts 5***									
	1) 8	2) 13	3) 8	4) 4	5) 14	6) 9	7) –	8) +	9) 9	10) 2
	Go On 9 – 6 = 3; 9 + 6 = 15									
	Mixed Facts 6									
	1) 6	2) 13	3) 10	4) 7	5) 12	6) 4	7) +	8) –	9) 7	10) 1
	Go On 13, 11, 5, 3; The numbers decrease by 2.									
p. 97	***Sums and Differences***									
	Answers will vary.									
p. 98	***What's the Sign? I***									
	1) –	2) +	3) +	4) –	5) –	6) –	7) +	8) +	9) +	10) –
	11) –	12) –	13) –	14) +						
p. 99	***What's the Sign? II***									
	1) +	2) –	3) +	4) –	5) –	6) –	7) +	8) –	9) +	10) –
	11) –	12) +	13) –	14) –						
p. 100	***Seeking Equations***									
	Answers will vary.									

p. 110	***What's in That Place? 1***				
	1) 25, 35	2) 37, 46, 48	3) 55	4) 72	5) 12, 15, 18
	6) 27, 37, 47	7) 45	8) 19	9) 49, 13	10) 10, 37
	Go On 47, 57, 67; Numbers increase by 10.				
	What's in That Place? 2				
	1) 16, 27	2) 33, 43, 44	3) 83	4) 67	5) 32, 36, 38
	6) 14, 34, 44	7) 56	8) 23	9) 11, 29	10) 47, 11
	Go On Answers will vary.				
p. 111	***What's in That Place? 3***				
	1) 22, 23	2) 37, 38, 39	3) 68	4) 74	5) 13, 14, 19
	6) 26, 46, 56	7) 34	8) 31	9) 38, 11	10) 11, 56
	Go On 38, 52, 61, 83				
	What's in That Place? 4				
	1) 27, 37	2) 44, 54, 55	3) 95	4) 52	5) 25, 27, 29
	6) 38, 48, 58	7) 48	8) 17	9) 28, 10	10) 12, 66
	Go On 31, 21, 11; The numbers decrease by 10.				
p. 112	***What's in That Place? 5***				
	1) 18, 29	2) 38, 47, 48	3) 77	4) 62	5) 44, 46, 48
	6) 13, 43, 53	7) 42	8) 36	9) 12, 48	10) 67, 13
	Go On Answers will vary.				
	What's in That Place? 6				
	1) 27, 28	2) 45, 46, 47	3) 86	4) 75	5) 54, 55, 57
	6) 29, 39, 59	7) 53	8) 17	9) 39, 12	10) 13, 58
	Go On 47, 68, 74, 82				

p. 119 *What Numbers are Missing? I*

1	2	3	4	5	6	7	8	9	10
11	12	13	14	15	16	17	18	19	20
21	22	23	24	25	26	27	28	29	30
31	32	33	34	35	36	37	38	39	40
41	42	43	44	45	46	47	48	49	50

11	12	13	14	15	16	17	18	19	20
21	22	23	24	25	26	27	28	29	30

31	32	33	34	35	36	37	38	39	40
41	42	43	44	45	46	47	48	49	50
51	52	53	54	55	56	57	58	59	60

p. 120 *What Numbers are Missing? II*

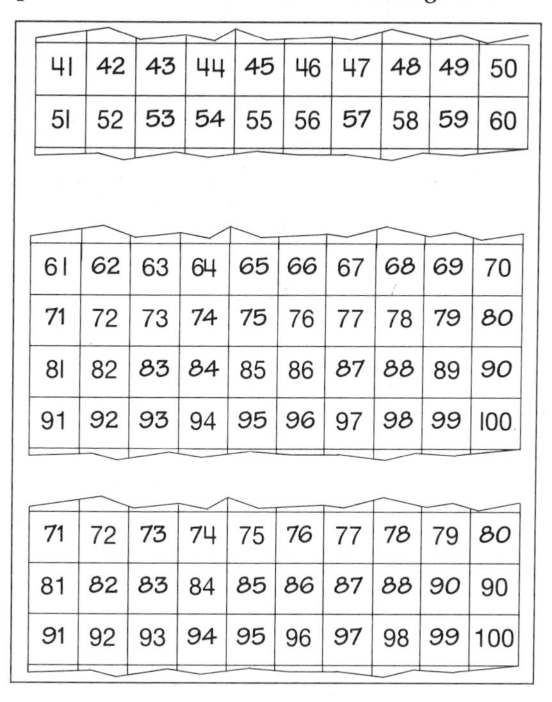

41	42	43	44	45	46	47	48	49	50
51	52	53	54	55	56	57	58	59	60

61	62	63	64	65	66	67	68	69	70
71	72	73	74	75	76	77	78	79	80
81	82	83	84	85	86	87	88	89	90
91	92	93	94	95	96	97	98	99	100

71	72	73	74	75	76	77	78	79	80
81	82	83	84	85	86	87	88	90	90
91	92	93	94	95	96	97	98	99	100

p. 121 *Fill in the Pieces I*

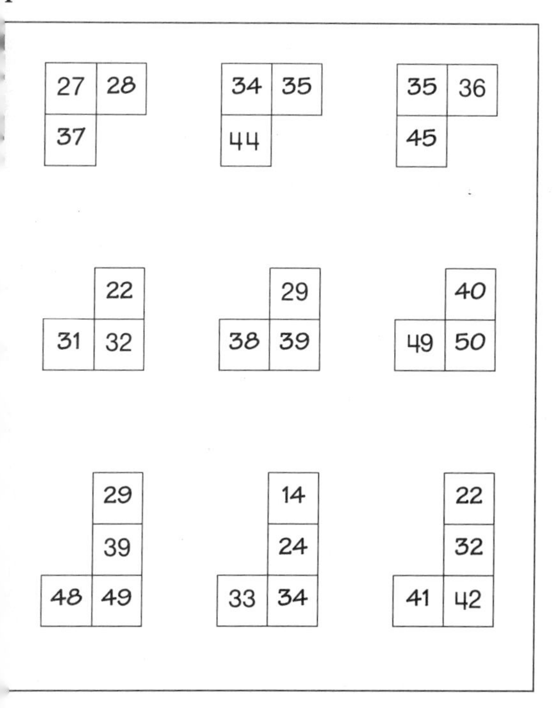

p. 122 *Fill in the Pieces II*

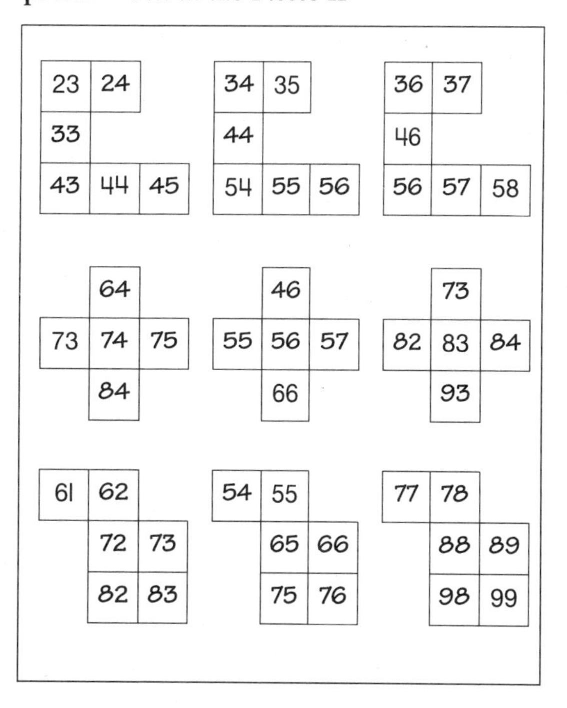

p. 123 *Smallest and Largest I*

1) 12, 42 2) 23, 53 3) 12, 52 4) 23, 43 5) 13, 53
6) 24, 54 7) 34, 54 8) 14, 54

p. 124 *Smallest and Largest II*

1) 46, 76; more 2) 68, 98; more 3) 35, 75; less 4) 28, 98; more 5) 56, 86; more
6) 36, 86; more 7) 45, 95; more 8) 24, 84; more